SEARCH OUT SCIENCE

BOOK 1
COMMUNICATION • STRUCTURES

Authors: **Mary Horn and Ann Orchard**

with Beryl Peters, Gill Gething,
Fen Marshall, Diane Ward

BBC Longman

Sources of light

Candles

Candles

Starting with the collection
- Light a candle and observe the flame.
- Light other candles and notice any similarities and differences.
- Light a candle in a dark place and compare with one in daylight.
- Does it matter what colour the wax is?
- Observe with your senses (smell, touch, look).
- Sort and group the collection in a variety of ways (shape, stability, type of wax, type of wick).

Exploration and investigation

Are all flames
— the same colour?
— the same size?

Put a cold saucer over the flame to see what happens.

Do the flames behave the same way in a draught?

What is this liquid? Collect some carefully in a tin lid.

Which candle will give the brightest light?

Put a piece of tissue paper in front of your face. How many layers before you cannot see the flame?

1 metre

candle on board or in sand tray.

Fair test
- Are you always the same distance away from the candle?
- Does it matter what colour the tissue is?

Can you devise a test to find out if...
- candles burn up at the same rate?
- two candles are twice as bright as one?
- it matters what the wick is made of?

Challenge
Can you make a candle clock?

[Note: you may need several candles all the same.]

Practical points
- Do not light a candle without asking first.
- Put lighted candles in a sand tray.
- Avoid putting them in busy places.
- Beware of trick candles which re-light when you blow them out.
- Beware of hot wax.

COMPARING THE CANDLES

CANDLE	wax colour	flame colour	flame size
1. thin red	red	yellow	3 cm
2. tall blue	blue		
3. round	yellow		

2

Sources of light

Torches

Torches

Starting with the collection
- Observe them (look, feel, listen . . .).
- What do they all have in common? What is essential for a torch?
- How are they different?
 (Look at switches, handles, the material they are made from, how you use them, what kind of beam they have.)

Exploration and investigation

Look at the beam:
- its shape and its pattern on the wall.

Look at the brightness.
- Is it good for reading under the bedclothes or looking for the cat in the dark?

See how far the beam travels.

How easy is it to switch on and off?

Can you find out:
- if you can see the torch beam through different fabrics?
- how the beam spreads out?

Class 3 put all their findings into a table and came to the following conclusions.

Can you devise a test to find out . . .
- which torch is best for sending signals?
- which torch is best for reading a map in the car in the dark?

Practical points
- You may need black material to help you make a dark place.
- The brightness of the torch will depend on the newness of the batteries – your results may vary.

Our class found out about torches

TORCH	ease of switching	stability	size	width of beam at 1m.
1. red	hard	unstable	will go in pocket	10 cm
2. large black	easy	stable	bulky	
3. black & yellow	simple	compact		

Our best torch is number 5 because it is easy to use in many ways, it has a wide beam and is the brightest.

3

Finding out about electricity

Circuits

Electrical items

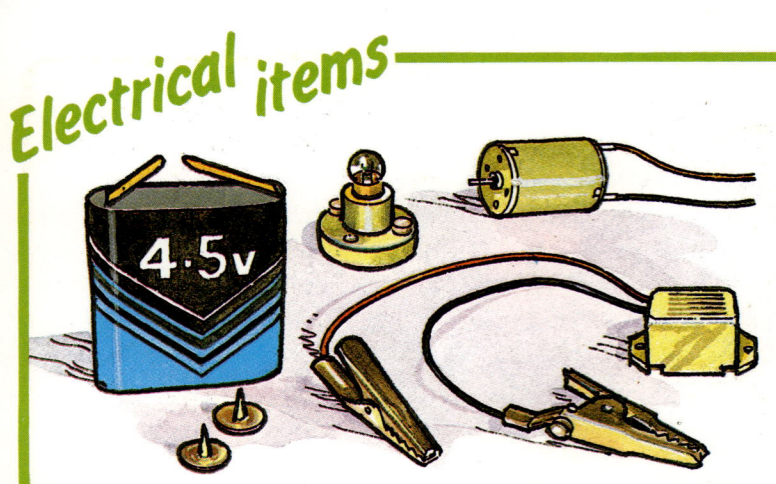

Finding out about electricity

The best way to find out is to have a go.
- Find out how to light the bulb.
- Are there several ways to do this?
- Can you make the buzzer work and the motor go?
- Make up a number of different circuits.
- Notice what happens if there is a break in the circuit.

Exploration and investigation

Does electricity pass through everything? What do you think?

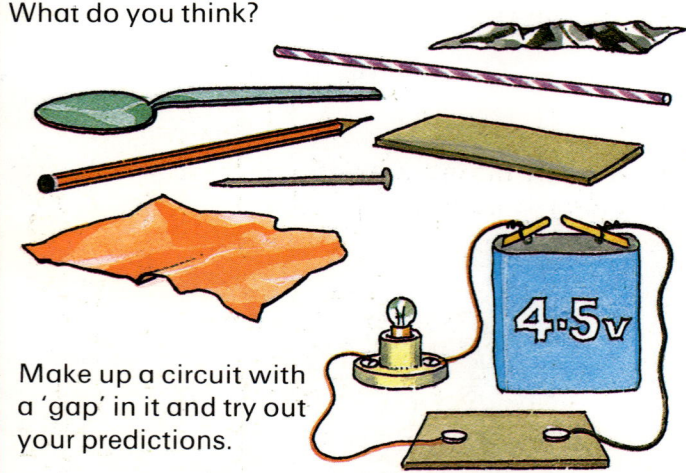

Make up a circuit with a 'gap' in it and try out your predictions.

Devise a way to find out what affects the brightness of the bulb.

- Use your circuit to try:
  - different lengths of wire
  - old/new battery
  - different 'gap fillers'.

- Now you have investigated your ideas, record what you think is happening.

Challenge 1

Can you make two bulbs light up at once?
- Do you notice anything about their brightness?
- Is there more than one way of doing this?

Challenge 2

Can you make a coloured light?
- Try different ways of making the light coloured.
  - Which way is most effective?
  - Are some colours better than others?

Challenge 3

Can you devise a system of sending messages using two different coloured lights which you can control?

Alison and I made up a circuit like this.

We used our circuit to test all sorts of things.

Never use mains electricity for your investigations. It is over 200 times stronger than a battery. That can be dangerous!

Using electricity to communicate

Switches

Switches

- paper clip
- drawing pins
- piece of foam to keep sides apart
- metal foil
- metal ball (or foil covered marble)
- metal foil

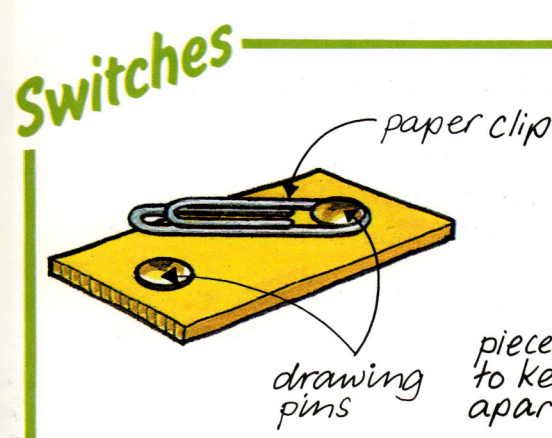

A simple switch like this lets you control the flow of electricity around your circuit.

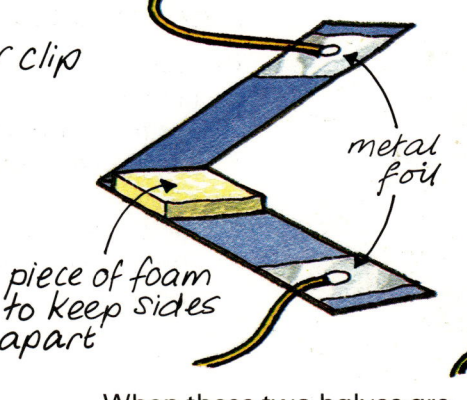

When these two halves are pressed together, the foil makes contact and electricity can flow.

A 'tilt' switch has a moving part which drops into a hole or gap to make contact, as the switch is tilted.

Using buzzers and lights...

to protect your valuables.

- Can you make a box which sounds an alarm when it is opened?

- Can you make a burglar alarm to warn you of intruders?

Challenge 1

Can you make an alarm which tells you when it is raining (so that you can shut the greenhouse)?

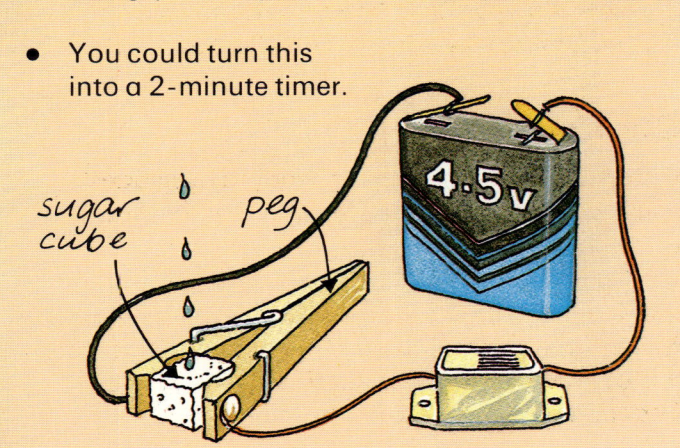

- You could turn this into a 2-minute timer.

sugar cube — peg — 4.5v

Challenge 2

Design an alarm which tells a *deaf* person that someone has stepped on the doormat.

5

How light behaves

Passing through materials

Seeing through

Starting with the collection
- Hold things up to the light.
- What do you notice?
  – How much light can you see?
  – How clear are the objects you see?
- Can you sort the collection according to what you noticed?

transparent	translucent	opaque
green bottle polythene bag	lunch box	lego brick

Exploration and investigation

- Put each object in front of the screen and shine a light on it.
- Look for patterns and look for shadows.
- Does everything make a shadow?
- Rotate the object and see what happens.
- Does the quality of the image or shadow change?

Does light pass through other materials?

- You may need to investigate the effects of:
  – more than one layer
  – very thin layers
  – colour in the material.

Challenge

Can you find out if materials change the way they let light through under different conditions?
- Paper is usually opaque. Rub it with oil, fat or candle wax and see if you can make it translucent.
- Perspex (or other clear plastic) is usually transparent. Rub the surface with sandpaper to make it translucent.
- Drop water on to the surface and see if you can make it transparent again.

Looking through liquids

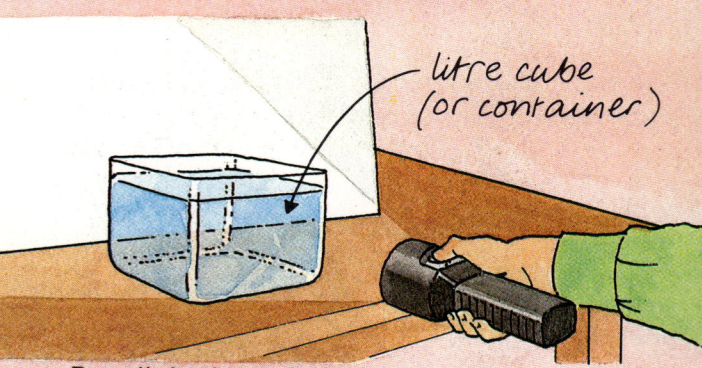

litre cube (or container)

- Pass light through the liquid.
- Try a little milk to make it translucent and watch the way the light beam passes through.
- Try colour in the water.
- Try other liquids.
- Record any pattern in your findings.

transparent – can see objects through
translucent – can see light through, but cannot see objects clearly
opaque – does not let light through

How light behaves

Forming shadows

Shadows

Finding out about shadows

Shadows are fascinating. Once you have found out about them you can use them to have lots of fun – perhaps put on your own shadow theatre for others.
- Cut out some shapes and hold them so that they cast a shadow.
- Does the shadow always touch the object?
- Cut out shapes from coloured plastic. What colour is the shadow?

Exploration and investigation

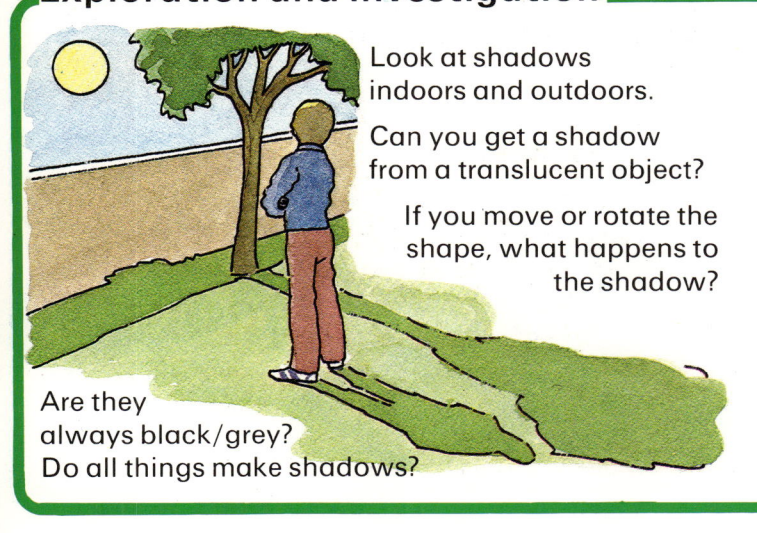

Look at shadows indoors and outdoors.

Can you get a shadow from a translucent object?

If you move or rotate the shape, what happens to the shadow?

Are they always black/grey?
Do all things make shadows?

Can you find out how shadows behave?

- Does the size of the shadow change if you move the object around?
- Can you find a pattern in what happens?
- Record your findings.

Watch what happens to the shadow of a stick (pole or gatepost) during a sunny day.

- Mark out where it is at different times.
- Mark out how long it is.
- You can use this to make your own sundial.
- In most sundials, the upright (gnomon) is a triangular shape. Can you find out why?

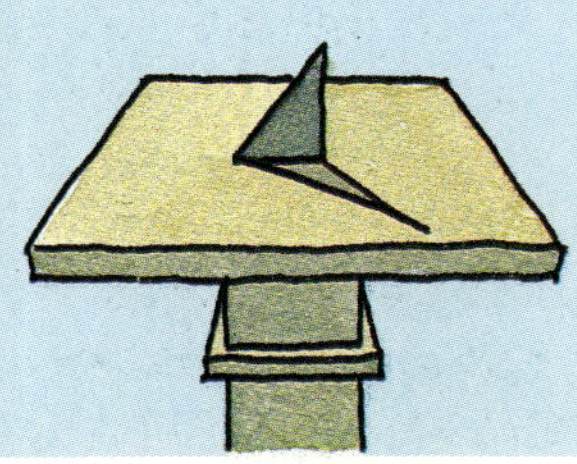

Mark out the times on the flat base. Is it reliable?

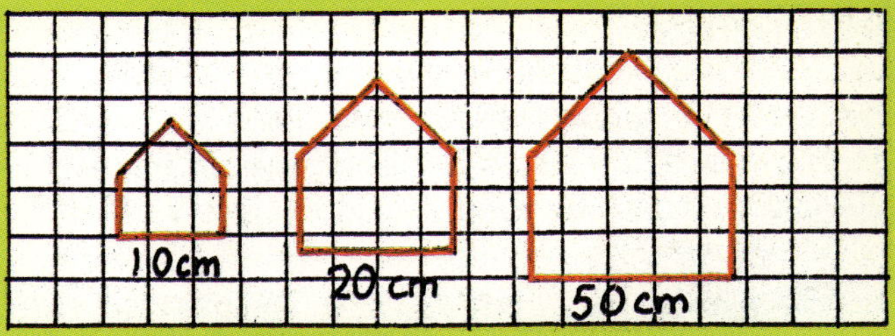

These are drawings of my shadows at different distances from the screen.

distance from the screen	area covered by shadow
10 cm	? sq cm
20 cm	? sq cm
50 cm	? sq cm

I found that there is a pattern. The further away the object gets from the screen, the larger the shadow.

How light behaves

Reflections

Shiny things

Starting with the collection
- Observe the shiny places and highlights.
- Can you make the highlights move?
- Which objects can you see your face in?
- Is your face always the same or is there any distortion?
- Can you predict what your face will look like in a surface?
- Record what you have found out.

Exploration and investigation

Use a small, flat mirror.
Can you see
– behind you?
– under your chair?
– under a table?
– over a wall?

Put the mirror under your chin and look up into the branches of a tree.

Can you find a leaf in the mirror, then find it again without the mirror?

Can you write your name when looking in a mirror?

Which letters/shapes stay the same when reflected – which ones change?

How many reflections can you see?

- Use plasticine to fix the mirrors so that they stand up.
- Look at the images in the two mirrors. How many can you see?

Does it matter where you put your head?

- Hinge two mirrors together with tape.
- Open and close them slowly and watch the number of images you can get. Is there a pattern in this?

Curved mirrors – do they make a difference?

Look at your face in both sides of a spoon. Record what you see.

Use a curved mirror (or a bendy one) to investigate a variety of images.

Can you predict what will happen? Can you explain the types of image?

Devise your own 'Hall of Mirrors'.

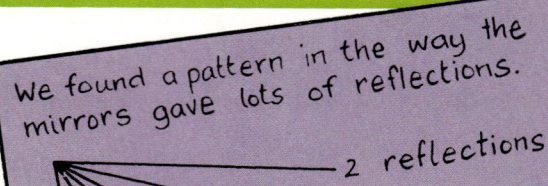

We found a pattern in the way the mirrors gave lots of reflections.

2 reflections
3
4
5
7
lots

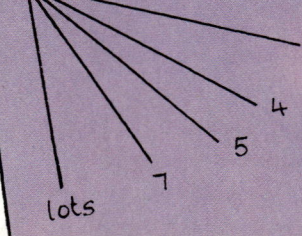

The closer the mirrors are together the more images you get.

- Find some ways of fixing mirrors so that you get multiple images. Try four mirrors in a square.
- What do you see? Is there a pattern in your findings?
- What would happen with five mirrors?

How light behaves

Beams of light

Beams of light

Light beams

Usually there is so much light around that it is difficult to see how it behaves.
- Sometimes we see shafts of light.
- If we want to study light it is better to make our own beams using a torch covered with black paper and a slit cut out.

> It is not always easy to see light beams in the middle of a classroom. It is better to work in a darker place – for example, a corner or a cupboard or under a table.

Exploration and investigation

Investigating how a light beam behaves

- Put some black paper over the torch.
- Cut a small slit in it so that a small, thin beam of light can shine through.
- Reflect this beam from a mirror and watch what happens.

- Move the torch.
- Does the light always travel in straight lines or can you make it move in a curved pathway?

Reflections

Investigate the way the reflected light behaves when bounced off a mirror.

Challenge

Can you design a mirror maze? First devise a road plan, a fantasy setting or a maze, and then place mirrors so that you can see round the objects.

This is how we investigated the reflections.

As we moved the torch beam, the reflection moved too. There is a regular pattern in the beam and the reflection. They are symmetrical.

9

How we receive light

Eyes

Eyes

iris
pupil

Looking at eyes
- Look at your own eyes in a mirror and look closely at a friend's eyes.
- Notice similarities and differences – watch the way the eye and lid move.
- Notice moisture and tears; colour, shape and size.
- How do eyes change when they have been closed for a few seconds?

Exploration and investigation

Devise a test to find out how well you and your friends can see
- close up
- at a distance.

5 metres

Do both eyes work equally well? How well do you see colours?

Devise an investigation to find out if right-handed people are 'right-eyed'.
- Most people find that one eye is more dominant than the other, in the same way that one hand dominates.
- Devise a test to determine 'eye domination'.

Fair test
- Can you test people with glasses too?
- Do the results depend on how long you have had your other eye shut?

Why do you need two eyes?
- Hold one finger out in front of you.
- Close one eye.
- Bring the other arm round fairly quickly and try to touch the two fingers together.

You need two eyes to judge distances and make your vision 3D.
- Look around you with one eye open and see what differences you notice.

Devise a way of finding out how good people are at judging distances with one eye. Is it something which gets better with practice?

What colour pattern do you see?

Some people cannot see the figure in the middle. They find it hard to discriminate between some colours.

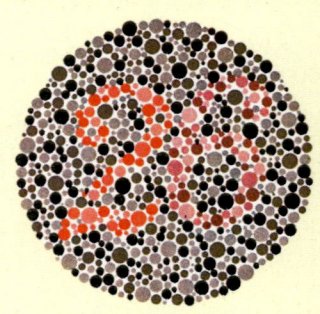

Eyes are very precious
- Never shine a torch directly into someone's eyes!
- Never look directly at the sun through binoculars or lenses!

10

How we receive light

Using lenses

Lenses & spectacles

Starting with the collection

- How clearly can you see through the lenses? Look at the picture.
- Do they make things look bigger or smaller?
- Can you sort the lenses into groups?
- Feel the lenses in each group. Do they have anything in common?
- Can you think of any other everyday uses of lenses?

Exploration and investigation

Helping eyes to see in more detail

Make a collection of things which magnify.

Feel the lenses and describe their shape.

What is the best position to hold the lens to get a clear picture?

How good is your magnifier?

- Does it magnify ×2 (make the object look twice as big) or ×3 (make the object three times as big)?
- Can you devise a way of finding out the magnification of a lens?

Lenses enable us to see a whole new world of the very small or fine detail. They also help us to see into the distance.

Looking through two lenses together

Combining lenses can give some very interesting effects.

Can you make things bigger or smaller?
Can you get things into focus?
Do some things appear upside down?

Vary the distance between the lenses.

Do not wear glasses which were made for other people for any length of time! It can damage your eyes.

Seeing in colour

Coloured things

Use coloured acetate sheet or colour paddles.

Seeing through colour

- Look through the coloured sheets at the different things around you.
- Notice the way the colours of some objects appear to change.
- Arrange a collection of coloured objects in front of you.
- Make yourself some coloured glasses and look through them.
- Can you predict what colours you will see?

Exploration and investigation

- Put a colour filter over a torch.
- Shine it on your collection of coloured objects.
- Record what you see.

- Make yourself a small-scale stage set or scene.
- Light it with torches using different coloured filters.

- Can you create different effects with light alone?

Looking for colours

- Blow bubbles, look at the colours and watch them change.
- Oil on a puddle gives a rainbow of colours.

- Shining a light through a glass prism can produce a lovely rainbow of colour. One particular position of the light produces a good rainbow. Can you find it?

Challenge

Can you make a vehicle and wire up lights for different purposes?

warning light (blue/yellow)
headlamps (white)
rear lights (red)

Super challenge

Can you make the warning light flash or revolve?

Seeing in colour

Now you see it, now you don't

Signs

Being seen
- Look at these signs. They are for warning and are easy to read. They stand out.
- What makes them stand out?
  – Is it the colour of the symbol?
  – Is it the colour of the background?
  – Is it the two together?
- Can you tell which is the best colour combination?

Exploration and investigation

Many things are designed to show up well in poor light or for safety at night.
- Make a collection of things you wear which help you to be seen.
- Can you devise a way to find out which is most effective
  – in the dark?
  – when a light shines on it?

Can you devise a test to:
- try different coloured backgrounds?

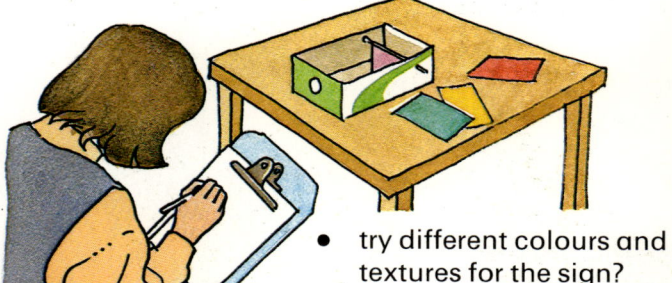

Fair test
Is one opinion enough?

- try different colours and textures for the sign?
- Do you get the same results in the dark?

Hiding away
- Cut out several moth shapes, like this one.

- Look for yourself at plants and animals which hide themselves in their backgrounds.
  – Look under stones and notice the colours of the creatures you find there.
  – Look amongst leaves and find how well caterpillars and other insects can hide.

- Colour them in to hide on:
  – a stone wall
  – the bark of a tree
  – a leafy plant
  – a pebble beach.

Can you make your own 'hidden display'?

13

Listening to sounds

Ears

How good are your ears?
- Do you use both ears equally?
- Can you devise a way to tell if you and your friends use both ears equally?
- If one ear is better, is it always the same ear?
- Are right-handed people 'right-eared'?

You may need to think about the direction of the sound and the distance away from the sound.

Exploration and investigation

Receiving sounds
- Many animals have large ears. Make yourself some large ears of different shapes.
- Test to see if they make any difference to how well you hear.
- Can you find a good shape and size?
- Did you make sure your testing was fair?

Which fabric makes the best insulator?
There are times when we want to concentrate and keep other sounds out.
- Devise a test to see which fabric makes the best ear muffs.

Listening post
- Choose three places around the school where you will hear different noises.
- Listen there for a short period of time, e.g. 5 minutes.

	outside classroom	by gate	dining room
people talking	|||| ||	||||		||||	
animal/bird noises		|||| ||||	||
cars/motor bikes			|||| ||	|||

- Record what you hear.
- Record in the same place on different days and at different times.
- What patterns are there in your data? Can you explain them?

Making the best ear muffs.
We recorded the number of layers of each fabric needed before we could not hear the sound.

fabric	cotton	news-paper	bubble packing	red felt
no. of layers	7	12	2	4

We could not decide how to test the cotton wool. We now think our test was not quite fair because some of the fabrics started off thicker than others.

Does it matter if:
- the fabric is closely packed in flat layers or loosely packed inside the ear muff?
- you listen to high sounds or low ones?

14

Listening to sounds

Tape recorder

Listening quiz

Listening carefully
We hear so many sounds and so much background noise each day that many people find it difficult to distinguish sounds from each other.
- Work with a partner.
- Put a collection of objects into a set of tins.
- Ask your partner to guess what is in the tin by listening to the sound when they shake it.
- How near was the guess? Did they guess the right material?

Exploration and investigation
- Using a tape recorder, gather sounds from around the school.
- Test other children. How many sounds do they recognise?
- Do some sounds seem easier to recognise than others?

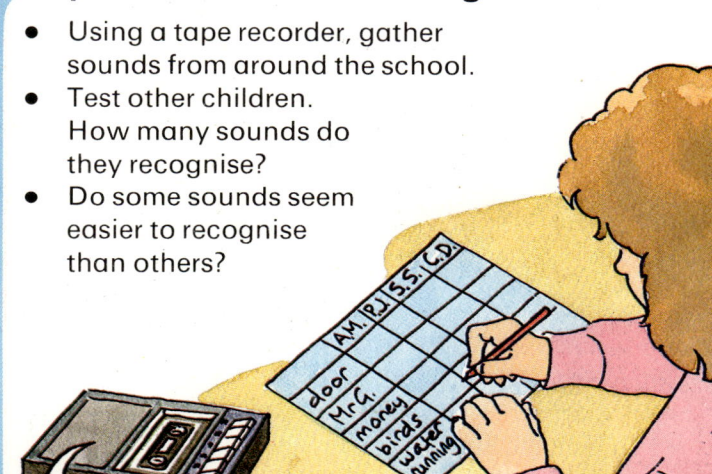

e.g. voices, typewriter, birdsong, water running, door shutting, recorder, money clinking.

Can you really hear a pin drop?
You may need to devise tests to find out if any of these have any effect:
— the direction you are facing
— the surface the pin drops on
— the way the pin is dropped
— how far away you are.

Fair test
Can you make sure people don't cheat — even unintentionally?

Sorting and grouping materials
We usually sort or group through similarities and differences we can see.
- Try hitting, dropping or tapping — then listening to see if you can distinguish between materials by the sounds they make.

Sound effects
These make a lot of difference to a story or play when you are listening.
- Make up some of your own sound effects for a story or poem.
- Remember things are not always what they seem and it is only the sound you need, e.g. coconut shells for horses' hoofs, and 'wobble' board for thunder.

Invite the audiometrician into your class and ask him or her to show you how people's hearing is tested.

Loud sounds can damage your hearing. Do not alter the volume when someone is wearing headphones!

15

Making sounds

Vibrations

Sound makers

Starting from the collection
- Make a sound – look carefully.
- Can you control the sound?
  – start it, stop it . . .
- Feel the instrument whilst the sound is going on. What do you notice?
- When the sound stops, what do you notice?

Exploration and investigation

Select two or three different types of instrument from the collection.

Create the sound and feel the vibration.

How long does the instrument vibrate?

Can you tune it?

Does it depend on
– how hard you hit it?
– what material it is made from?

Can you make the sound in different ways?
- bang it
- blow it
- pluck it
- tap it

What is vibrating?

Investigating a vibration

Use a drum or instrument with a skin.

Put some small pieces of lightweight material on top (peas or rice or expanded polystyrene). Tap it with different beaters and notice how the skin vibrates.
- Does it matter
  – how hard you hit it?
  – where you hit it?

Use a tuning fork
- Set it vibrating by tapping the open end.
- Dip the end into water carefully whilst it is still vibrating.

When we speak we make the sound by creating a vibration in our voice-boxes.

Put your fingers over your voice-box as you talk and feel the vibration.

Electrical vibrations

Sometimes you can create vibrations electrically. Feel a buzzer whilst it is making a noise. Can you see the vibration too?

You can see and feel the vibration of speakers.

Nowadays many vibrations can be created electrically.

16

Making sounds

Banging

Percussion

Sounds can be high or low. This is called **pitch**.

Sounds can be loud or quiet. This is called **volume**.

Some sounds go well together and make **music**.

You should be able to put together a varied collection of pitched and unpitched instruments from around the school.

Group them into those which you bang, pluck or blow, so that you can study the way they work.

Starting with your collection

- Find out what kinds of sounds are made from different materials.
- Use different beaters and change the type of head (metal, wooden, brush).
- Try wrapping material around the solid beaters. Does it make any difference?
- What produces different volume?

Creating pitched sounds from percussion

flower pots (clay is best)

metal pipes
- try solid and hollow pipe
- try dowel rod or other wood

- Try out some ideas of your own.
- How do percussion instruments vary the pitch? Do you always need a set?

Steel drums are beaten out to include several 'tuned' patches on each drum. Can you find out how this is done?

Investigating ways to increase or dampen the volume

For sounds to reach your ear the air has to carry the vibration.

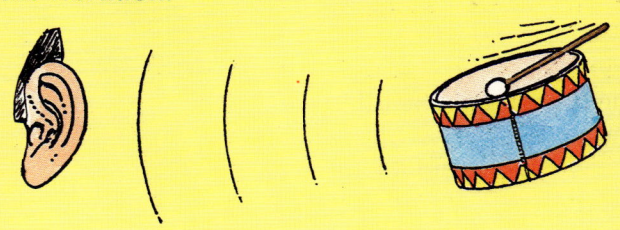

You will need to find ways of making the most of the vibrations. 
- Try standing the drum on different surfaces:
  – a wooden table
  – a metal tray.
- Try inside a cardboard box and on top of an upside-down box.
- Try banging the triangle
  – whilst you hold it.
  – holding it on a string.

17

Making sounds

Plucking

Strings & things

Starting with the collection
- What happens to the string when you pluck it? Look closely and see if all strings do the same thing.
- Can you alter the pitch of the note – making it higher or lower?
- Does gently plucking make any difference to the pitch of the note? If not, what does it change?
- Look carefully at the strings on instruments. What do you notice? Are they all the same?

Exploration and investigation

Hold one end of a plastic ruler firmly over the edge of a table. Twang the other end.
- Watch it vibrate.
- Does it produce a note?
- Can you vary the pitch of the note?

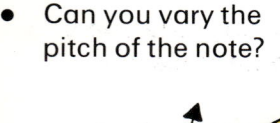

Vary:
- the length over the table edge
- the force of the twang
- the way you hold the other end.

Try other rulers or strips of material.

Investigate the way a rubber band vibrates

There are two things which can vary to change the pitch of the note:

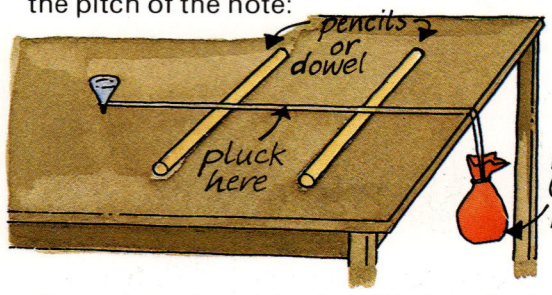

- how tight the rubber band is (tension), and
- the length of the part vibrating.
- Vary the tension using weights.
- Vary the length of the string by moving the pencils (or dowel).

Investigate the way you can change the pitch

- Can you predict what is going to happen?
- Is there a pattern in your findings?
- Record what you did and what you found out.

Challenge

Make a stringed instrument of your own which will give you a wide range of notes.

Can you investigate the design of your instrument and perhaps modify it to make it give a louder sound?

Do different strings produce different types of note?
Try wire, string, fishing line or dental floss.

Rubber band investigation

Jane's group varied the weight on the end of the rubber band.

Karen's group varied the distance between the pencils.

Mark's group tried different rubber bands.

We compared our findings.

Making sounds

Blowing

Wind instruments

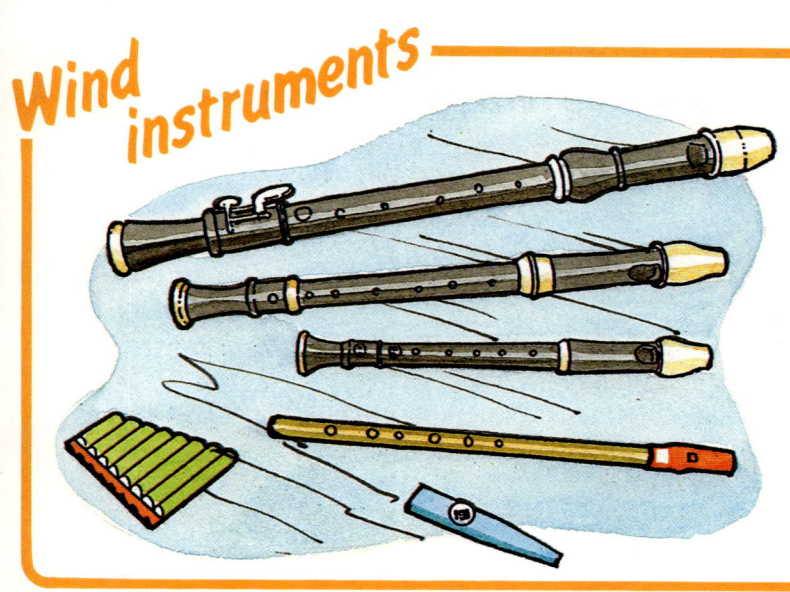

Starting with the collection
- Try blowing the objects in the group and see how you can make the sound.
- What is vibrating? Can you see or feel anything?
- Does it make any difference if you blow hard or gently?
- What seems to change the pitch?

Exploration and investigation

Collect a series of bottles the same size.

- Blow gently *across* the top and see if you can make a sound.
- What happens to the sound if you pour water into the bottle?
- If you tap the bottles, do you get a sound at the same pitch?

Investigating the sound from the bottles
- Try standing the bottles on different surfaces – metal, carpet, wood.
- Does the liquid inside make any difference?
- Does the material of the bottle make any difference?
- Can you make your own musical instrument from a bottle collection?

Challenge

Can you make your own 'pipe' out of an artstraw?

- Try cutting the end like this and blow using it as a 'reed'.
- Can you make several with high and low notes?
- Can you feel the vibration through your lips?

Whistling of the wind

On a windy day the wind whistles through holes and gaps, setting up vibrations which produce high sounds. You can do this too.

1. Release the air from a balloon through a thin hole in the neck.
2. Whirl a cardboard tube round above your head.

Sending sounds over distances

Tubes

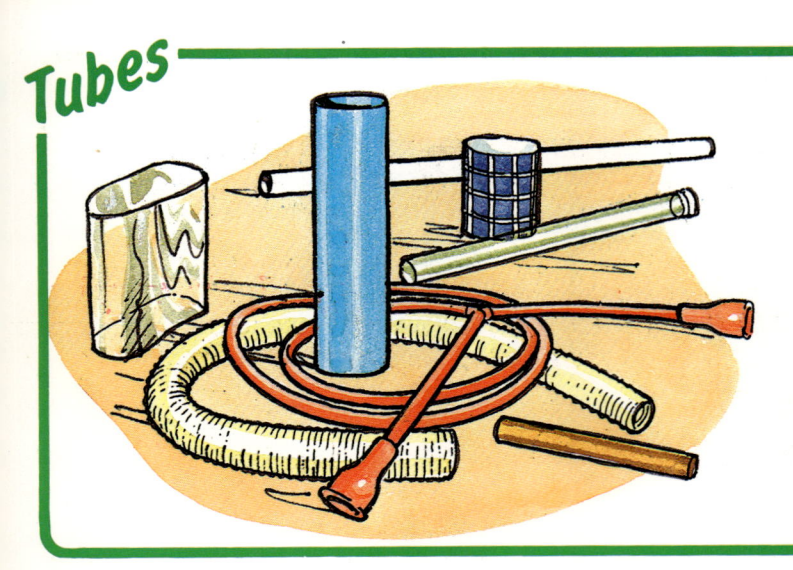

Starting with the collection
Make a collection of tubes, hoses or cardboard rolls joined together.
- Listen to sounds through the tubes.
- Try with straight tubes and bent ones. Does it make any difference?
- Try looking through the tubes bent and straight. Can you compare the way light and sound behave?
- Does the material of the tube affect the sound?
- Do the length or diameter affect the sound?

Exploration and investigation

Collect some large containers.
- Talk into them. What do you notice?
- How does your voice alter?
- Can you work out what is happening to the sound?
- Can you predict where else you will be able to make this echo effect around the school?
- Were you right?

Design the best megaphone.

This will need to be tested over quite a long distance.

- You will need to find out if the shape of the megaphone or the material make any difference.
- Does it matter what the weather conditions are like outside?

Does sound travel through different materials?

- Do you hear approaching feet better with your ear to the floor?
- Can you hear sounds through heating pipes from one room to the next? 
- Can you make sounds travel along a wall?

Make your own stethoscope

You will need a piece of tubing – preferably with a funnel in either end. (Why do you think funnels are a good idea? Try with and without them.)

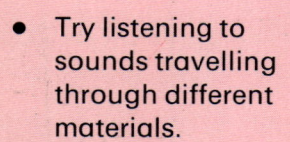

- Try listening to sounds travelling through different materials.
- Listen to a friend
  – tapping the bark of a tree
  – tapping the school pipes.

Sending sounds over distances

'Sound shadows'

- If you put a screen between a light source and yourself you cannot see it. Try the same thing with sound.
- Can you hear round corners or on another side of a wall?
- Does the wind make any difference?
- Can you hear high and low sound equally well?

Exploration and investigation

Make a 'string telephone' and explore ways of sending messages.

- Use a long piece of string and two containers.
- Fix the string.

Actually, this does not need to be string. (It does not work the same way as a telephone.)

- Speak into one container – listen through the other.

Investigate:
- A variety of strings – which is best?
- A variety of containers – does size or material matter?
- How far can you make your 'telephone' work?
- Can you make a 'telephone' which works round corners?

Same receivers – different strings.

	Tel. 1	Tel. 2	Tel. 3
1 metre	✓	✓	✓
2 metre	✓	✗	✓
3 metre	✗	✗	✓

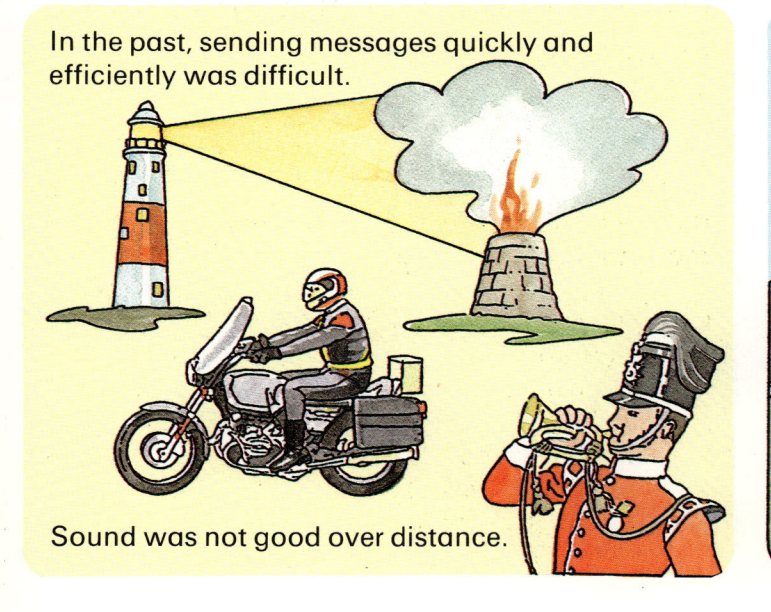

In the past, sending messages quickly and efficiently was difficult.

Sound was not good over distance.

Today sound vibrations can be transmitted by electrical means and now we use this as one of the best ways of communicating.

Challenges

These challenges are to set you thinking and doing. There is no right answer. For each one you will need some knowledge which you can combine with your own ideas. You will soon find yourself setting your own challenges. If your first ideas do not work out, talk to others – including your teacher.
— Happy problem solving!

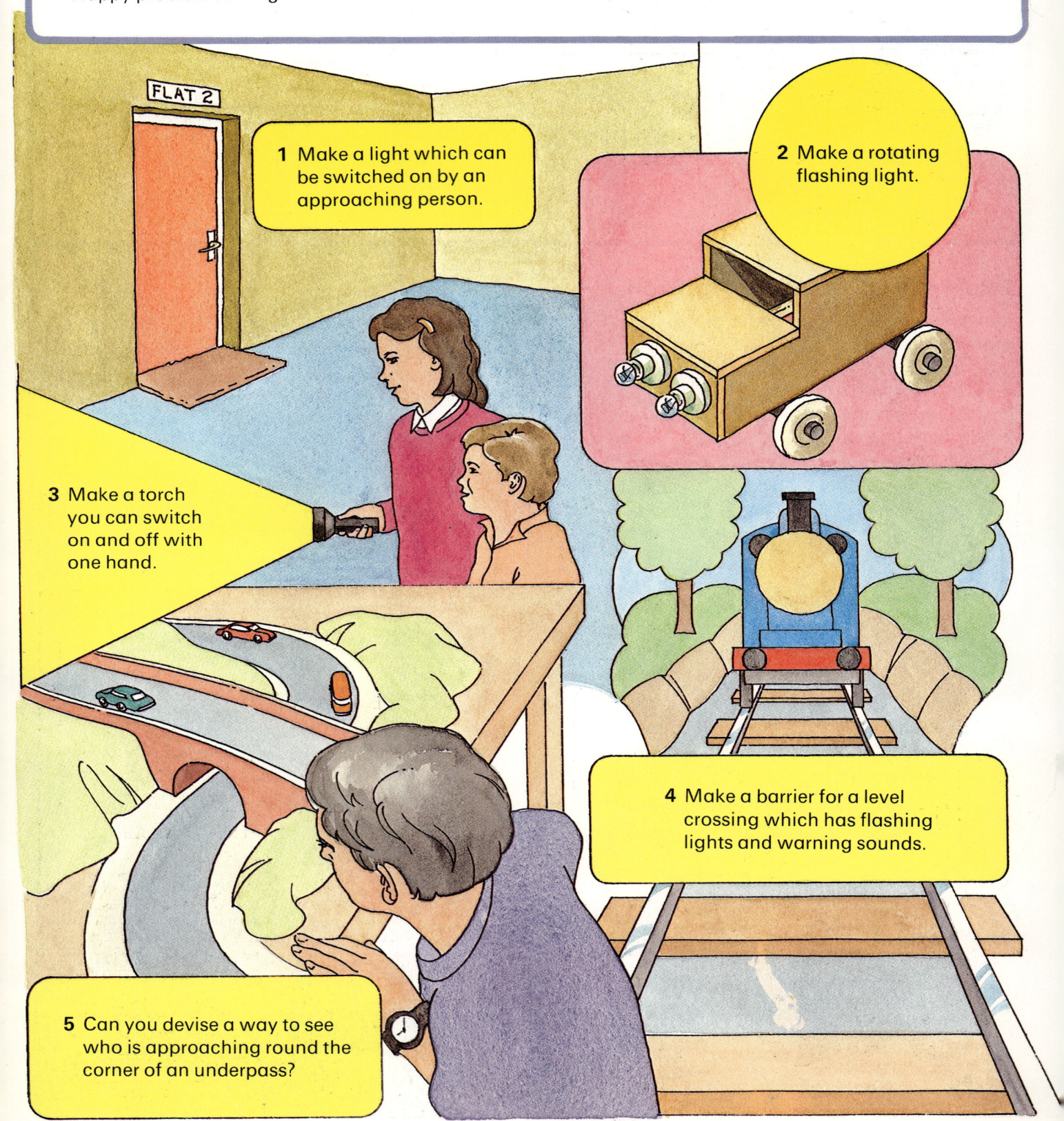

1 Make a light which can be switched on by an approaching person.

2 Make a rotating flashing light.

3 Make a torch you can switch on and off with one hand.

4 Make a barrier for a level crossing which has flashing lights and warning sounds.

5 Can you devise a way to see who is approaching round the corner of an underpass?

Challenges

6 The enemy have red visors. Can you devise signs they will not see?

7 You need to conserve candles. What factors do you need to consider? Test them, then design the best candle holder.

8 You need to leave messages only your friends can read. (You can tell them to carry a mirror.)

9 Make a trip switch to protect you while you sleep.

10 Can you devise a way to make the telephone ring sound louder to the old lady in the next room?

Structures

To understand structures, you need to start by looking at forces.

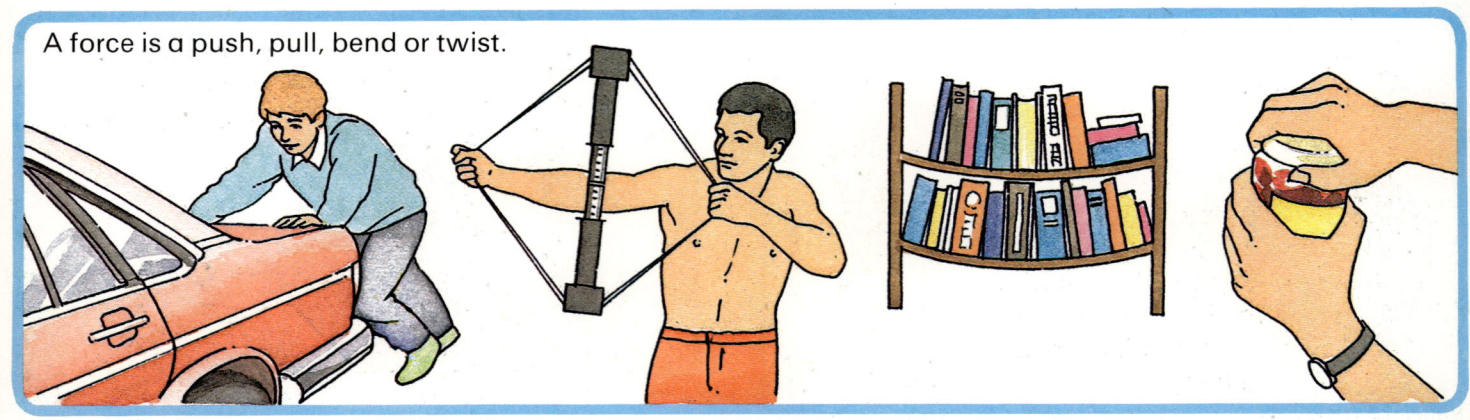

A force is a push, pull, bend or twist.

A force can set things moving . . . or stop them.

Forces sometimes prevent things moving or hold them steady to prevent them falling over.

Forces come in all sizes – from very big to very small.

24

Structures

Some forces have special names

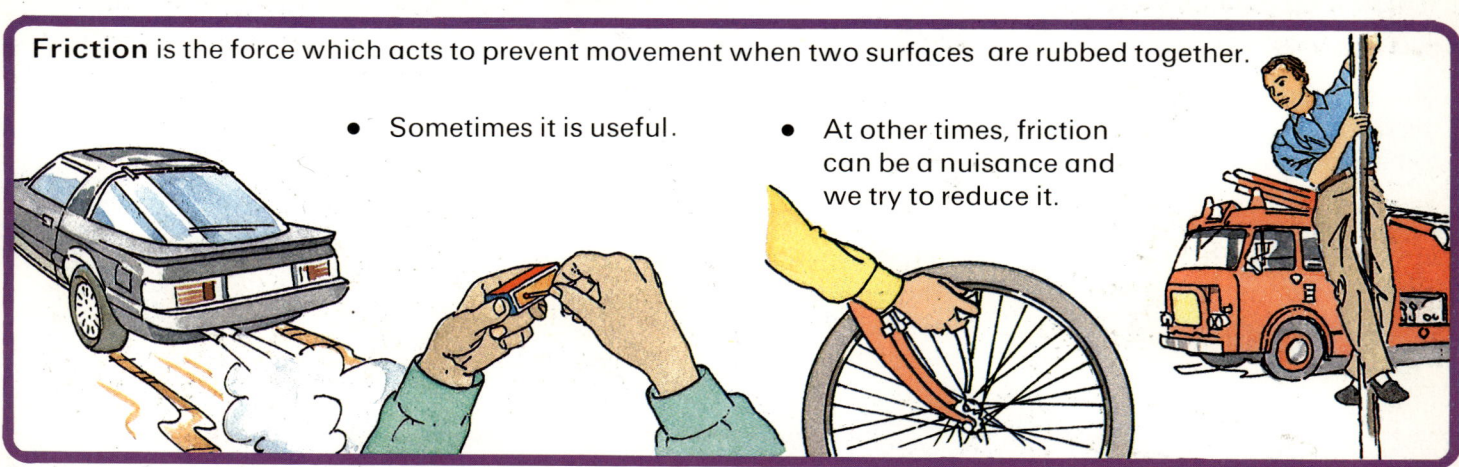

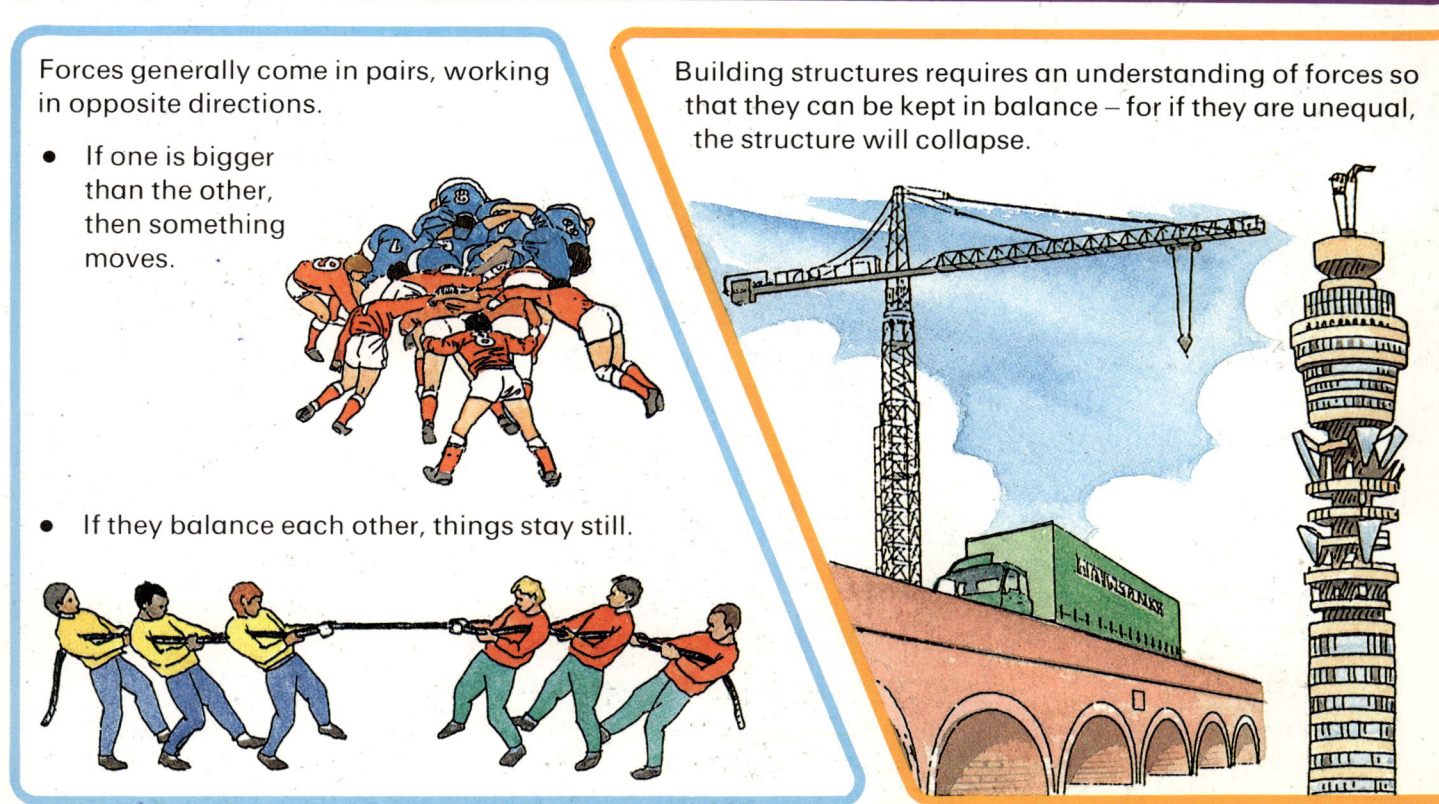

Gravity is the pull of the Earth (or any planet). Anything which moves away from the Earth's surface is pulled back by gravity.

It pulls down rain from clouds, fruit from trees... or you from walls!

Friction is the force which acts to prevent movement when two surfaces are rubbed together.

- Sometimes it is useful.
- At other times, friction can be a nuisance and we try to reduce it.

Forces generally come in pairs, working in opposite directions.

- If one is bigger than the other, then something moves.
- If they balance each other, things stay still.

Building structures requires an understanding of forces so that they can be kept in balance – for if they are unequal, the structure will collapse.

25

Forces all around you

Bicycles

Bicycles

Starting with a bicycle
- You will need to bring one or more bicycles into school.
- You are going to consider the forces involved in:
  - starting
  - stopping
  - going slowly
  - turning corners
  - going down a bumpy road.

Exploration and investigation

Investigating the forces associated with a bicycle

Although you may have ridden a bicycle many times, thinking about the forces involved may be new to you. Record what you do and the forces you feel when riding, or when turning, stopping or starting.
- What do you feel?
- Where do you push?
- Where do you twist?
- What happens as a result of your force?
- Do you use different forces at different times?
- What is the effect of the force? Is the pull or push passed on to other parts of the bicycle?

Forces involved with my bicycle

kind of force	part of bicycle
push	on pedals when starting & moving
push	on saddle when you sit down
pull	when you squeeze the brake lever

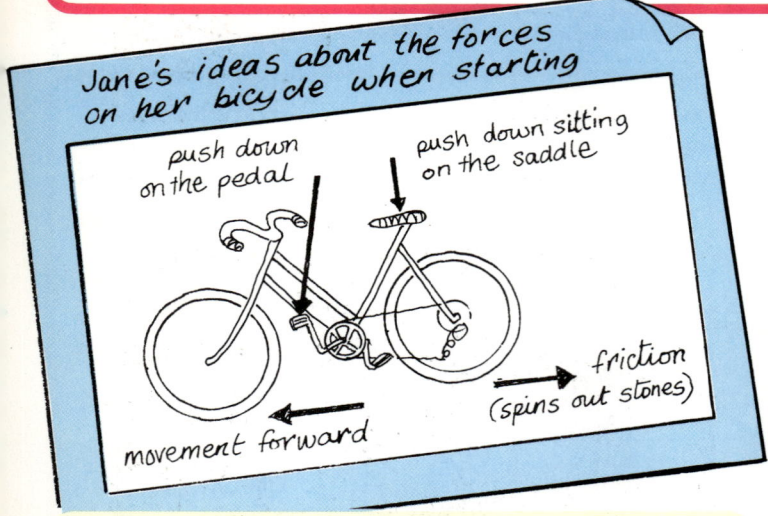

Jane's ideas about the forces on her bicycle when starting

- push down on the pedal
- push down sitting on the saddle
- friction (spins out stones)
- movement forward

Simon recorded his ideas like this.

First I squeezed the brake lever.
↓
This pulled the brake cable.
↓
This moved the brake blocks on to the tyre.
↓

Pumping up the tyres
- This involves using forces.
- You *push* the pump in and you can *feel* the resistance. The air in the tyre is also pushing against the tyre wall.
- Try this and record your findings.

Problem

A person with a load of shopping carries it on the handlebars.

How does this affect the other forces?

Is this the best place to carry the weight? If not, where would you suggest and why?

Forces all around you

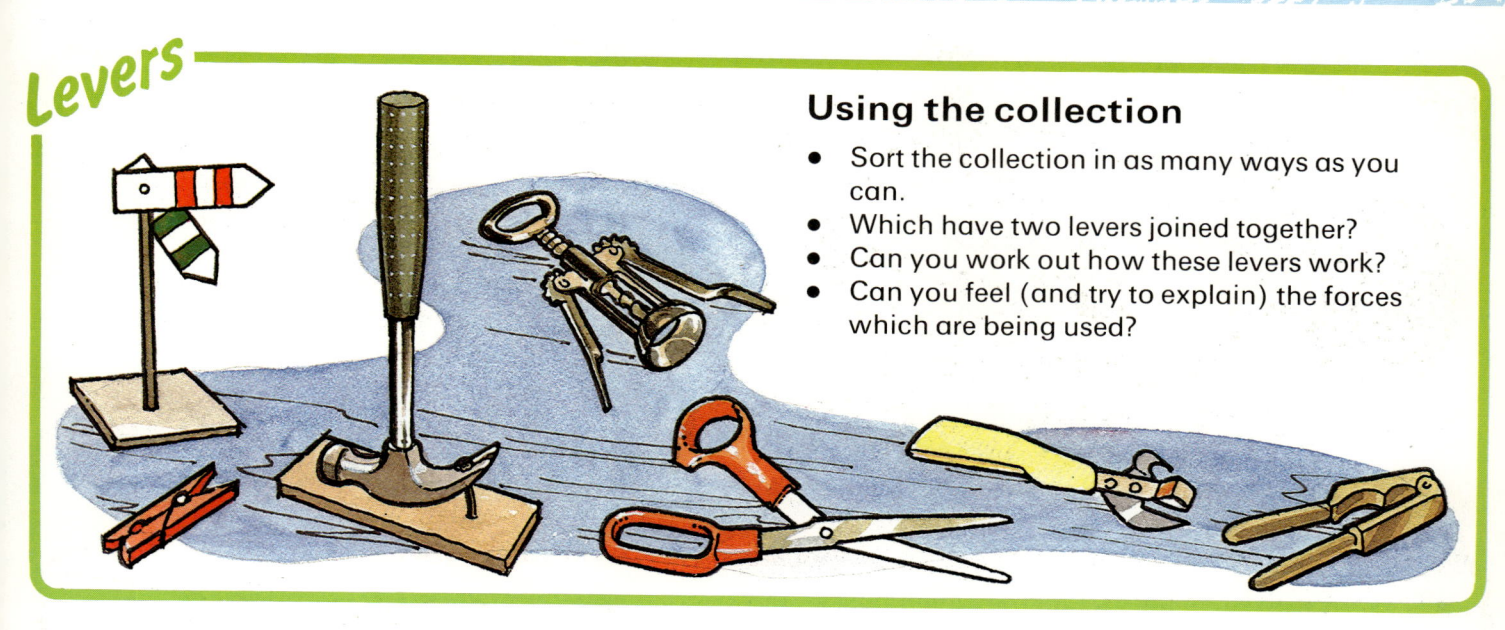

Levers

Levers

Using the collection
- Sort the collection in as many ways as you can.
- Which have two levers joined together?
- Can you work out how these levers work?
- Can you feel (and try to explain) the forces which are being used?

Exploration and investigation

Forces acting in one direction can produce a movement in another direction. This is often done using levers.

- You can investigate the way levers work using
  - strips of card
  - mapping pins
  - soft board.

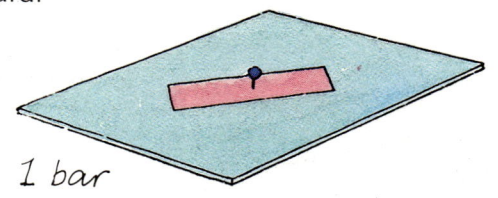

1 bar

1. Use one strip of card and a pin. Investigate the way the card moves with the pin in different positions along it.

2. Use two strips of card joined at the ends (with a pin), and investigate how the cards move with the pin in different positions. What difference does it make when cards are joined in the middle?

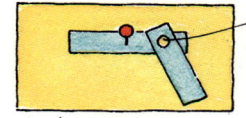

brass split pin

2 bars

Levers with more than two bars

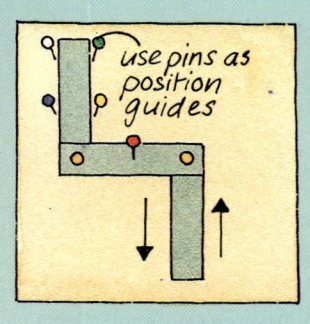

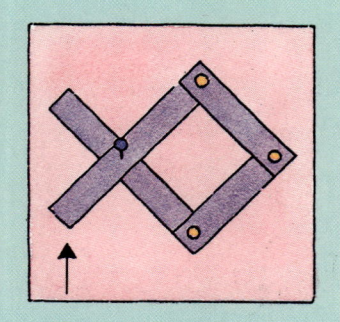

use pins as position guides

Explore systems with more than two bars of card.
- Use extra pins as position guides.
- Can you make a horizontal movement create a vertical one?
- Try this with card and construction kits.

Can you make a moving picture with levers?

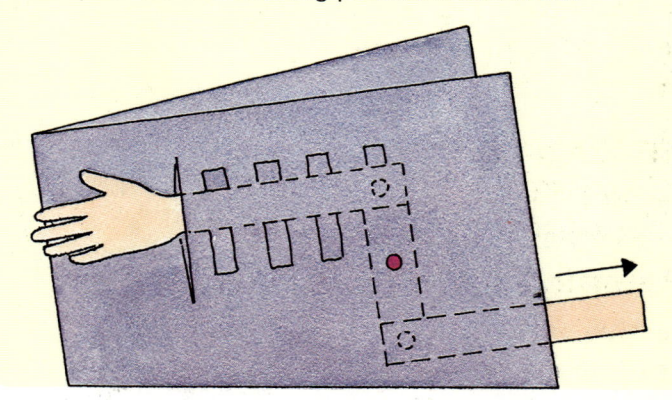

Challenge
Can you make a lever turn a wheel?
- Try this with card, and with construction kits.

Balancing forces

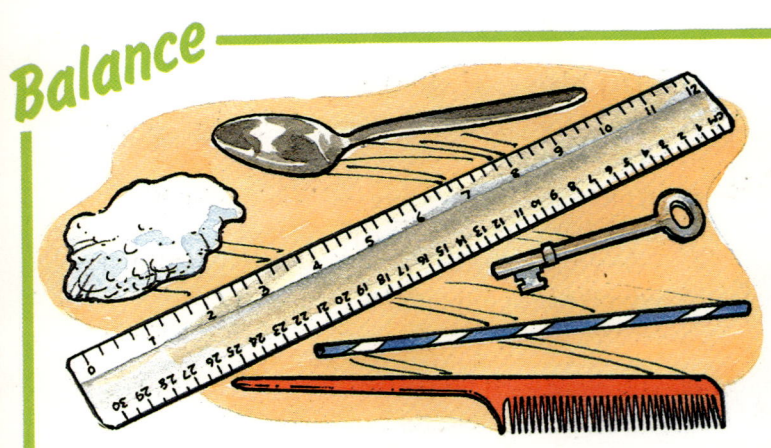

Balance

Starting with the collection
Make a collection of objects – some symmetrical, some asymmetric – made from a variety of materials.
- Push each one slowly to the edge of the table and note where it topples.
- Mark the point with a chalk line.
- Can you predict at which point it will topple over?
- Does it matter which way round it is?

- Take a small tray (or plastic lid) and balance it on a brick.
- Load it with small bricks, blocks or centicubes.
- Vary the position of the load. What do you notice about the balance?
- Can you predict what will happen as the small bricks shift their positions?

Exploration and investigation
Suspend two of your objects from a wire, like this. (Use florists' wire or thin wood strips.)

- Where do you need to hang similar objects to make a balance?
- Try a longer horizontal main strip. What happens?
- Record where the balancing point is.

Investigating a balance for small weights
- Can you use this device, or one like it, to measure some very small things? Try a leaf, seed, pin, 1 square cm paper.

wire counter weight *pin through approximate point of balance*

Make a mobile
- You will need:
  – materials for the horizontals (e.g. wire or thin wood)
  – something to hang or weigh.
- Start from the bottom.
- Consider how many horizontals you could use.
- Make your first mobile with identical hangings.

- Try a series of paper sheets – use the whole sheet but fold/crumble/cut and bend etc.
- Does this balance? Why?
- Vary the number of hangings.
- Use the mobile to illustrate a food chain, or a family tree or a set of birds.
- Hang the mobile over a radiator. What happens?

Balancing

Stability

Stability

Your own stability
- When you are doing PE, try different ways of balancing.
- What do you notice about balancing on one leg with arms in or arms out?
- As you walk along the thin edge of a form, is it easier with arms in or out?
- Stand on one leg and lean over forwards to pick something off the floor.
- How many different ways can you do this without falling over?

Exploration and investigation

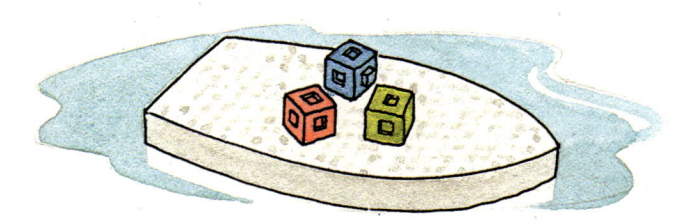

Investigate ways of making a boat more stable when you load it
- Use a piece of polystyrene, as before.
- Load it until it topples.
- Slot a keel in place.
- Does it make it any more stable?
- Does a mast make it more stable?

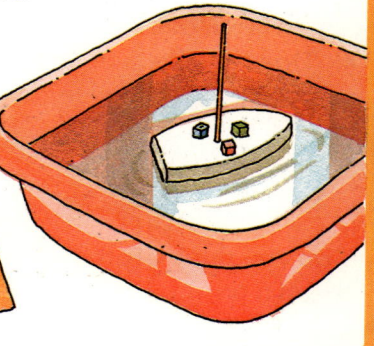

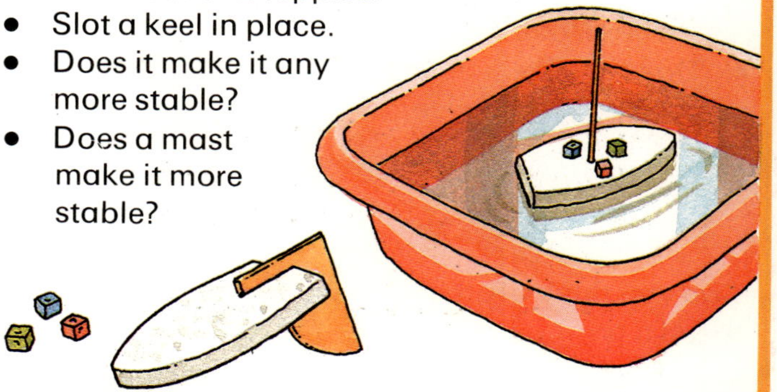

- Float a polystyrene 'boat' in a bowl.
- Load it with weights such as centicubes or unifix cubes.
- How many will it hold before tipping them off?
- Does it matter how you load them? Try them spread out and piled up.
- Record you results.

Investigating stability of bottles
- Make a collection of everyday bottles.
- Stand them on a plank and tilt it.

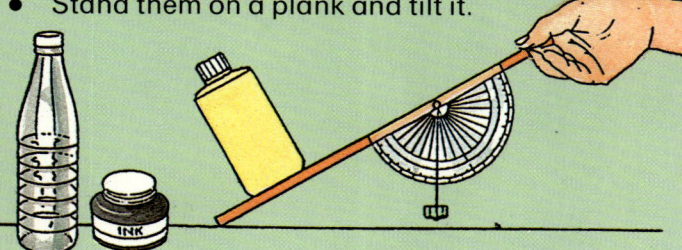

- Make a note of the angle you tilt it up to before it topples.
- Half fill the bottle with water and compare the stability.
- Is the same bottle most stable when it is half full of water?
- Is there a pattern in your findings?

THINK! Do you need to test more than once?

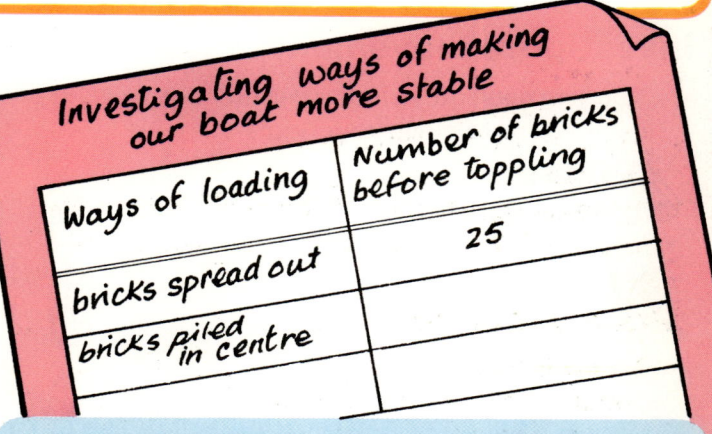

Investigating ways of making our boat more stable

Ways of loading	Number of bricks before toppling
bricks spread out	25
bricks piled in centre	

Challenge
Make your own 'wobbly' person.
- One segment of an eggbox makes a good base.
- How much plasticine is needed to make it stable?
- Where does the plasticine need to be put?

Resisting forces

Friction

Friction

Sometimes friction works *for* us.   Sometimes it works *against* us.

Exploration and investigation

- Find a screw top jar and do it up tightly.
- Now try to undo it with
  – a soapy hand
  – a dry clean hand
  – a wet hand
  – a cloth over your hand.
- Does it make a difference?
- What do you notice?

Can you think of other occasions when you need to get a grip?

Investigate conditions for slipping

- Use a brick or similar object.
- Pull the brick along the table.
- Record the reading on the Newton meter.
- Change the surface of the brick (bind some sandpaper around it) or change the surface on which you pull (floor, carpet, paper).

Investigate movement in water

- You will need guttering or a water bath, or somewhere to move your boats.

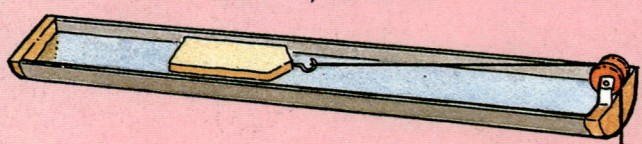

- If you do not already have a set of shaped boats, make some out of balsa wood.
- Attach a piece of string and pull the shaped boats through the water.
- How well do they travel?
- Can you compare them?
- Which shapes are best?
- Can you predict what kind of shape moves most easily through water?

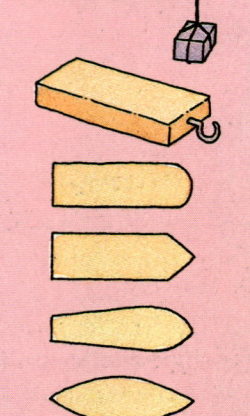

Pauline pulled a brick along different surfaces

surface of brick	surface underneath	Newton meter reading
brick	table	4.2
sand paper	table	5.3
brick	sugar paper	4.9

Challenge

Can you make a water wheel?

A water wheel uses moving water to turn it. Look for places where there might be resistance to the movement. How can you overcome those? Think first about some of the forces involved – it will help your design and planning.

Resisting forces

In air and water

Wind

Moving through air
- Take a large sheet of card or board. Go outside and walk towards the wind, holding the board so that it faces into the wind. (If it is not very windy, run with the board between you.)
- Try the same thing with the board end-on.
- Can you feel the difference?

Exploration and investigation

- Take identical sheets of paper and drop them – flat/crumpled/edge first/folded into a dart etc.
- What differences do you notice in the way they fall?
- Can you find any pattern which would help you to predict what will happen?
- How can you time the fall?
- Try different qualities of paper and see what happens.

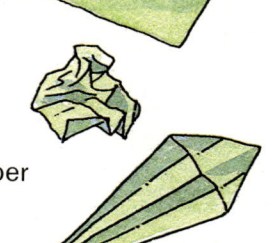

Investigate what makes a good parachute
You will need to try:
- different papers
- different sizes
- different weights
- a hole in the canopy
- different lengths of string.
- Which drops fastest/slowest?

All this data could be put on to a computer program, such as OUR FACTS or GRASS. Put in your own categories and ask your own questions. Look for patterns.
- What combination gives the best parachute?

Moving on land, in water or in the air

You need to push up because gravity pulls down.

You can use the force of wind and water to do work for you...

or you can make your object streamlined to minimise the effects of the force.

You need to oil or grease surfaces which rub together to help them move easily.

If you want to minimise the resistance, lubricate the surface and streamline the shape.

31

Which materials will you use?

Containers

Containers

Starting with the collection
- Observe the containers – look and feel.
- Sort the collection in as many ways as you can. Think about:
  - material
  - flexibility
  - transparency.
- Are lids and handles made from the same material?
- Record your groupings.

Exploration and investigation

Using your collection of containers
- Which ones would be best for holding water? Devise a test.
- Which ones would be best for holding a hot drink? Devise a test.
- Is the same material best for keeping an ice cream from melting?

Challenge
Can you make an 'egg carrier' – a way of packaging one egg so that it can be carried safely?
You will need to consider which materials to use – they will need to be light yet strong.

Investigating materials
Cut a series of strips of different materials.

- Devise a way of finding out which of them
  - is strongest?
  - are waterproof?
  - conduct electricity?
- Can any of these materials be stretched?
- Which ones will stand up to wear and tear?

THINK! Do all your samples need to be the same size?

We tested a range of materials and put down our findings.

No.	material	strong	water-proof	conduct elect.	wear & tear	floats
1	newspaper	x	x	x	x	x
2	polythene	✓	✓	x	✓	x
3	J-cloth	x	x	x	✓	✓
4	iron bar	✓	✓	✓	x	x

Compromises!
One of the problems of choosing the 'right material for the job' is that it may not have the perfect combination of properties.

Material	Advantages	Disadvantages
expanded polystyrene	light-weight good insulator cheap floats	breaks easily
iron	strong	rusts

Materials

The same or different?

Wood

Collect some wooden items and some off-cuts of wood. Look for similarities and differences.

Observing your collection

Look at the grain and smell the fresh wood.
- Is all wood equally hard? If not, how can you find out and compare them?
  Devise a test.
- Does all wood float?
  Does some wood float better than the others? Can you devise a test to find out?
- How strong is a lolly stick?
  How do you make a comparison?
  Are two lolly sticks twice as strong as one?

Plastics

There is a very wide group of materials belonging to the plastics family.
- What properties do you normally associate with plastics? What similarities are there?

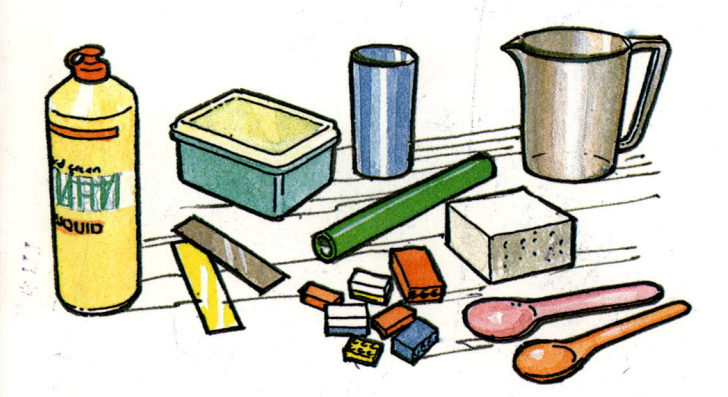

- Consider the following when applied to a group of plastics:
  - transparency
  - pliability (bendiness or flexibility)
  - strength when thick and thin
  - waterproof.
- Collect a group of plastic beakers. Which is the best value for money?
- Consider a range of properties:
  - stability when full
  - insulation (Can you pick it up when full of hot liquid?)
  - durability (How easily do they break or damage?)
  - price.

Can you decide which is the best?

Metals

There is a wide range of metals, though many of them are not in common use: from iron and tin which are quite common, to mercury (a liquid) and sodium (too reactive to be useful on its own).

- Collect together a range of items made from different metals.
- Make a chart to give three similarities and three differences.
- Test as many metals as you can to see if they are all:
  - magnetic?
  - electrical conductors?
  - strong?
  - shiny?
- Metals can often be made into thin wires – you cannot do this with wood.
- Compare woods, plastics and metals. Make your own tests. Do not assume anything!

33

Fastening and joining

Fastenings

Starting with the collection
- Observe the collection, using all the appropriate senses.
- Sort the collection in as many ways as possible. Look at:
  – the materials
  – what it is fixed to
  – type of fixing
  – what forces it has to withstand.

Exploration and investigation

Joining paper and card
- Collect a range of different types of paper and card.
- Collect a range of fastenings which could be used with paper.
- Are all fastenings equally suitable? Under what circumstances are they used to best advantage?
- Which fastenings would hold a box on to a collage like the one below?
- Does it matter how heavy the box is?

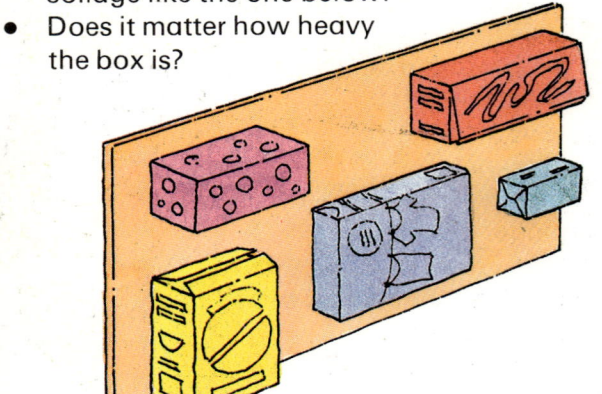

Investigating carrier bags
- Make a collection of carrier bags of different types – as varied as possible (e.g. different materials, sizes, handles, styles).
- What do you think makes a good carrier bag?
- Make a list of the variables which you think have an effect on the quality of the bag.
- Test your ideas. Make sure you are fair.
- Load the bag.
- Note where it gives way.

We tested our CARRIER BAGS with 'WOW' cat food tins. Here are our results.

Bag	Where it gave away	Number of tins
big bag	handle	10
white plastic	bottom seam	4

Fastenings often work best in one direction. Try separating them by pulling in different directions and noting the effect.

[Note: disasters often happen because a structure is hit by a force from an unexpected direction.]

Always load the bag near the floor.

Challenge
Make your own carrier bag.
- Fix the handles in different ways.
- Test them to find out which fastening is best.

Fastening and joining

Joining wood

- Collect some pieces of wood or board all the same size.
- How many ways can you join them together?
- Can you now test which way of joining is the strongest?

Exploration and investigation

- Nail pairs of wood pieces together with only a small overlap.
- Try different forces on your samples.

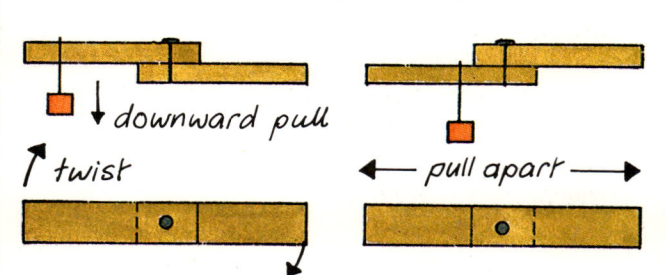

↓ downward pull
↑ twist
← pull apart →

- Are all joinings equally effective when pulled in different directions?
- Try other ways of joining (screw, string, glue). Do you get the same result each time?

Investigate the best glue for the job

There are many different glues and many materials to be joined.

- Make a series of small models. Use different materials and different glues.
- Test the joins.
- Record your data.

Our class has been model making this week. We used many different materials and tried out 7 glues. These are our results.

Materials	Glue	10 min	30 min	1 day
paper/paper	PVA	poor	good	good
paper/wood	copydex			

You could also put your data on to a database such as OUR FACTS or GRASS.

- Can you say which is the best glue?

Look at your collection of fastenings again.

- In which direction do they work best?
- Think carefully about the forces involved.
- Record your findings in diagrams, showing the forces.

strong (vertical direction)
pulls out easily (horizontal direction)

NOTICES
FIRST TEAM FOOTBALL
There will be a practice on Friday 16th

Investigate knots

Many groups of young people learn to tie a variety of knots.

- Test to see which ones:
  – lie flat
  – are strong
  – don't slip
  – tie thick string to thin.
- Test them with string, nylon fishing line, rope or cotton.
- Are the knots equally strong when the string is wet?

Making things stronger

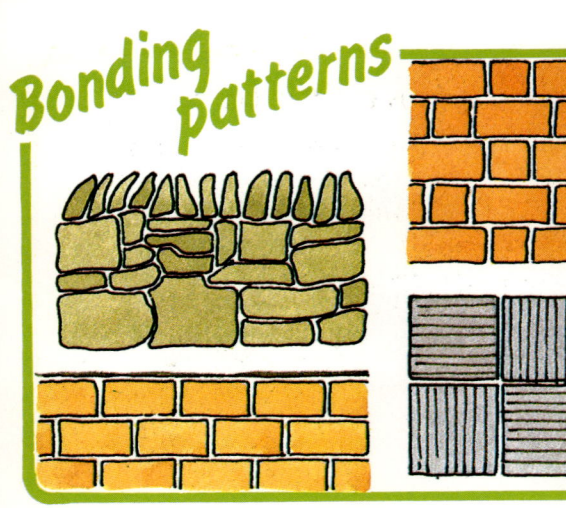

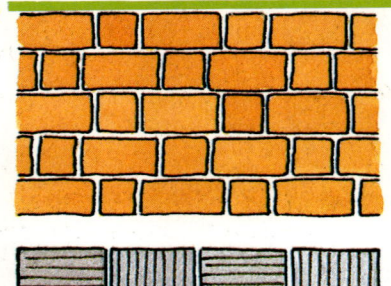

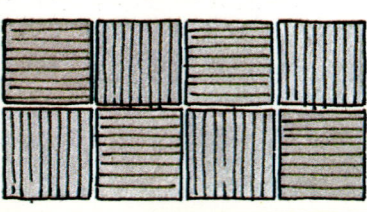

Arrangements

Bonding patterns

- Look at the way walls are built.
- Make a survey of the materials used and the arrangements of 'bricks'.
- You could add to your survey when on a visit or using pictures.
- Is there a pattern or series of patterns associated with one kind of material?

Exploration and investigation

Sometimes the problems of the large forces in the world do not show up on small models.

What happens to a large wall when a high wind blows?

Try it with a model.

- How can you support this wall to avoid it blowing over?

Investigate the strength of different wall bondings

- Build a series of model walls with bricks or lego blocks.
- Can you devise a test to find out which one is the strongest arrangement?
- How can you ensure that the force they all received is the same?
- Does it matter where the force hits the wall?

Making mortar and testing the mixtures for strength

Mortar is used to join bricks or blocks. It is made from cement, sand and water.

- Try mixing some mortar. Use different amounts of sand and cement.

Measures (spoons)			
Sand	cement	water	strength
3	3	1	SN
3	2	1	
3	4	2	

- Which mix sets quickest?
- What else do you notice?

- One group tried dropping a heavy weight on to their samples.
- Another group hammered nails into their samples.

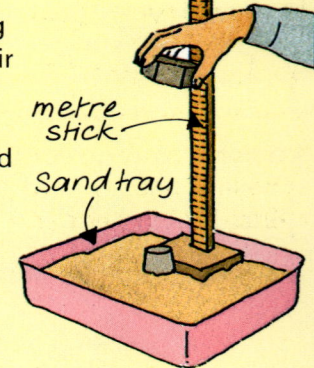

metre stick
sand tray

- Try these and some ideas of your own. Make your test fair.

matchboxes and plastic cups make good moulds

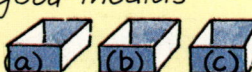

Protect your eyes from cement dust and mortar chips when testing.

36

Making things stronger

Arrangements

Ropes, cables & threads

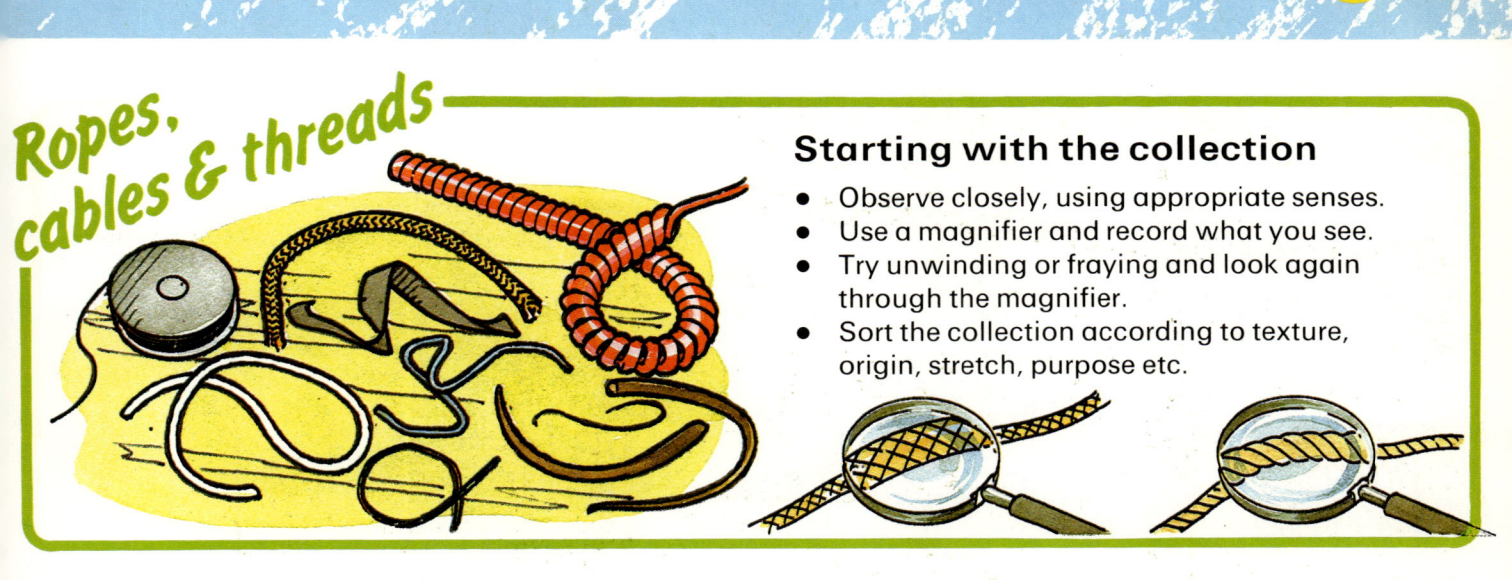

Starting with the collection
- Observe closely, using appropriate senses.
- Use a magnifier and record what you see.
- Try unwinding or fraying and look again through the magnifier.
- Sort the collection according to texture, origin, stretch, purpose etc.

Exploration and investigation

Exploring twisted fibres
There are many different kinds of thread made by twisting fibres together. Yarn for knitting and weaving must be strong yet flexible.

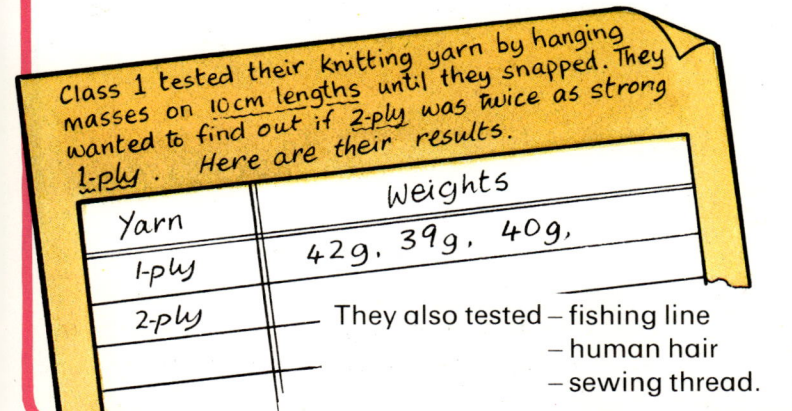

Class 1 tested their knitting yarn by hanging masses on 10cm lengths until they snapped. They wanted to find out if 2-ply was twice as strong 1-ply. Here are their results.

Yarn	Weights
1-ply	42g, 39g, 40g,
2-ply	

They also tested
– fishing line
– human hair
– sewing thread.

Investigate the strength of 10 cm of knitting yarn
- Try a range of yarns, from 1-ply to 4-ply.
- Does the yarn contain nylon? Does that matter?
- Devise a fair test and record your findings.

Living stems are reinforced with bundles and fibres that provide support.

Explore a collection of stems.

rhubarb   celery   twig

- Feel the ridges.
- Look closely at the ends of stems to see the structures.
- Try splitting these stems vertically and see if you can see the vertical fibres.

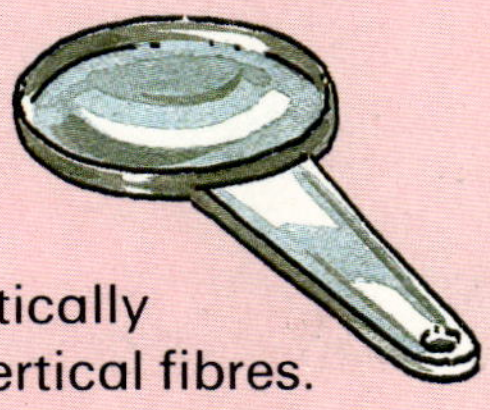

Testing the strength of bundles

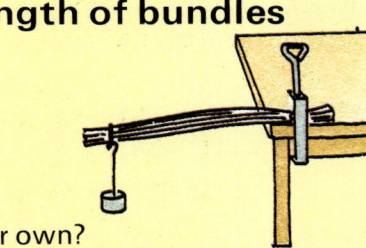

Here are two ways.
- Can you devise your own?
- Do twice as many strands make it twice as strong?

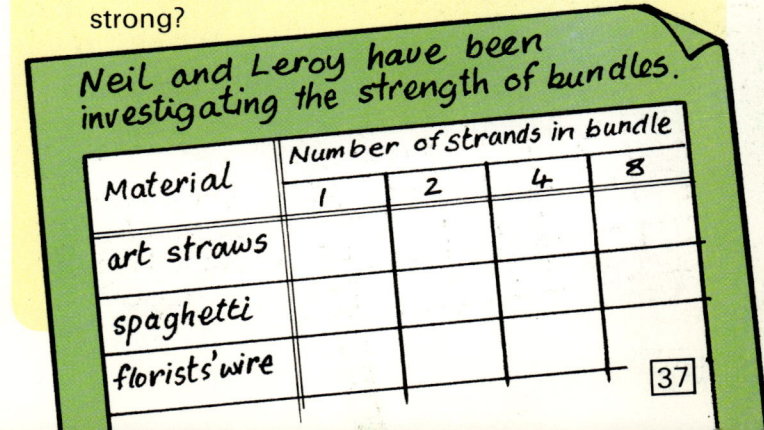

Neil and Leroy have been investigating the strength of bundles.

Material	Number of strands in bundle			
	1	2	4	8
art straws				
spaghetti				
florists' wire				

Keep weights away from toes.
Work close to the ground.

Making things stronger

Shape matters

Light & strong

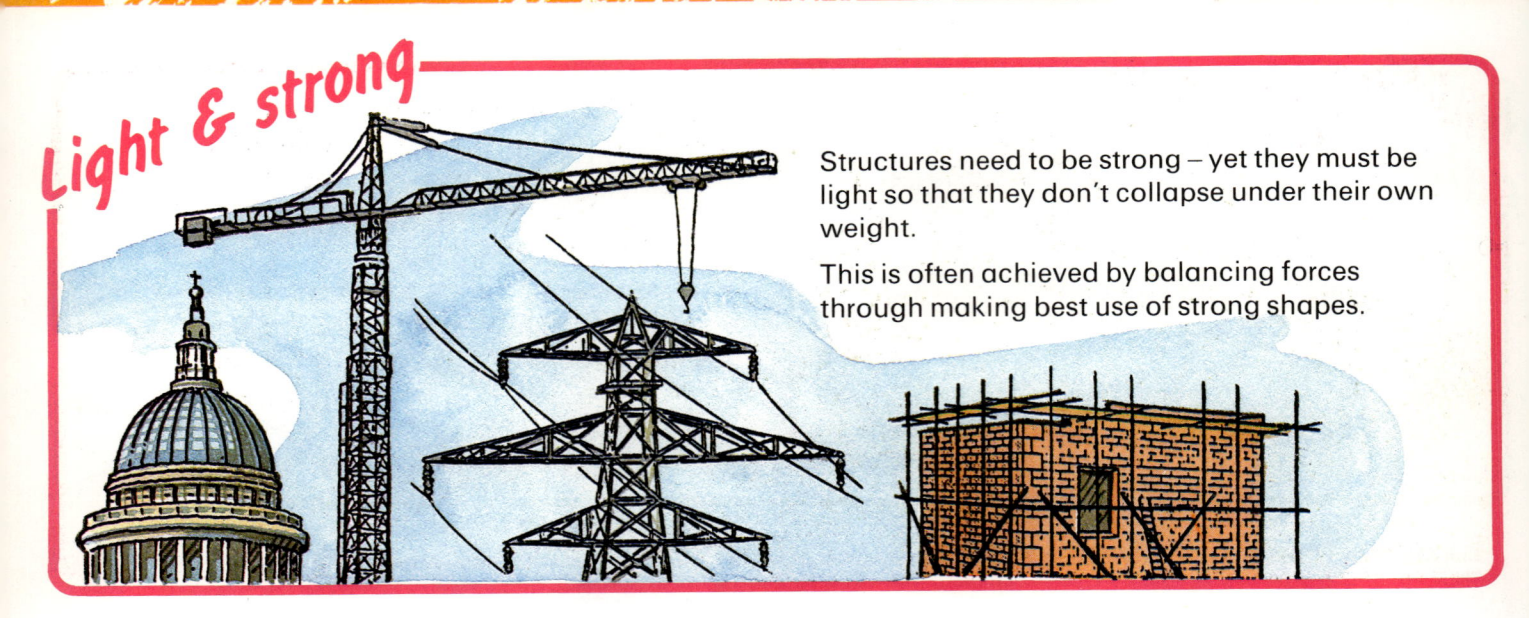

Structures need to be strong – yet they must be light so that they don't collapse under their own weight.

This is often achieved by balancing forces through making best use of strong shapes.

Exploration and investigation

Spanning the gap

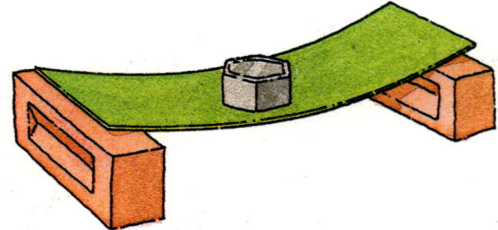

Use a piece of thin card to span a 50-cm gap.
- What load will it take before it collapses?
- Does it matter how or where you place the load?

Investigating ways of making the piece of card stronger

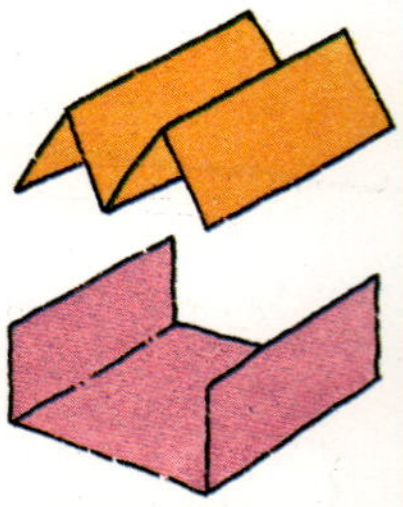

- Take several pieces of card all the same size and fold them in different ways.
- Try loading them and seeing what weight they take now.
- Where are the weakest points?
- Can you strengthen them?

We have been folding card in different ways to compare their strength. These are our findings.

Strength of shapes

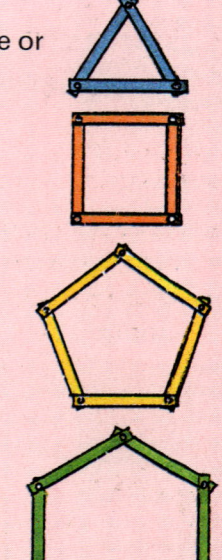

Use meccano type strips, corriflute or strips of card fixed with split pins.
- Make a shape with three sides.
- Make a shape with four sides, another with five sides and another with six sides.
- Can you deform the shape?
- For building structures it is important that the shapes do not move if a force pushes on them.
- Make these shapes rigid by putting the minimum number of struts across them.

Challenge

Use triangles to build the longest horizontal structure you can.
You will need card sheet and strips of card or corriflute.

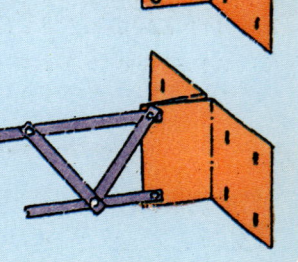

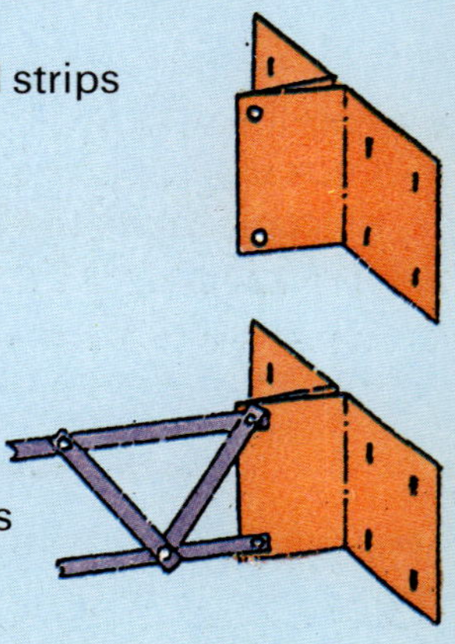

1. Staple a piece of folded card to the wall.
2. Use strips of card or corriflute to build out a strong structure, extending as far as possible. Secure the strips with split pins.

38

Making things stronger

Supporting loads

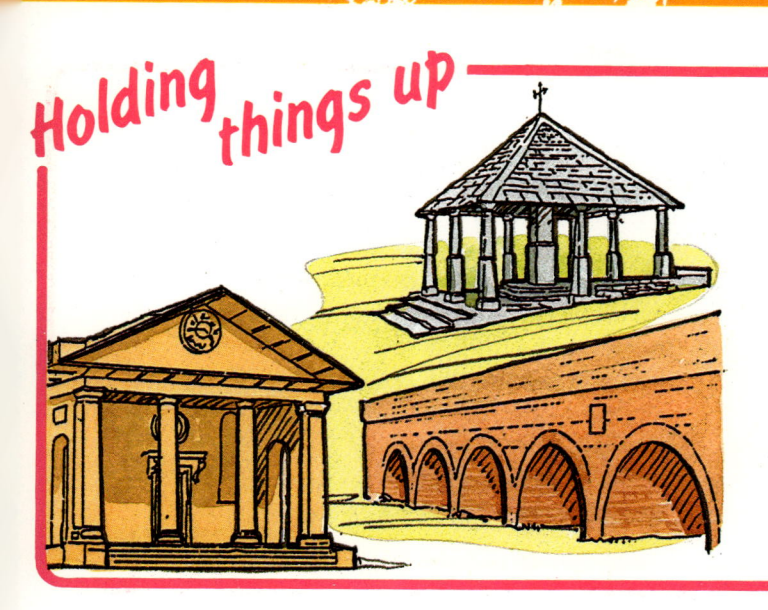

Holding things up

Supporting loads is a very important consideration for designers and engineers. There are two aspects of the support:
– the weight of the structure itself
– the load it has to carry.

Remember, too, that the problems and forces involved when making small models are not the same when they are scaled up to full size. Consider:
– speading the load/forces
– using lightweight materials
– using open structures (like triangles).

Exploration and investigation

Loads are often supported on PILLARS

Make some pillars of different cross-sections and compare the loads they will take before collapsing.

- Try tubes with different diameters.
- Does the type of paper make a lot of difference?

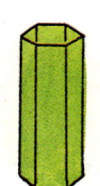

use paper

Investigating the strength of pillars

- Try different shapes.
- Try different heights of the same shape. Is a short fat pillar better than a tall thin one?
- What makes a good pillar?

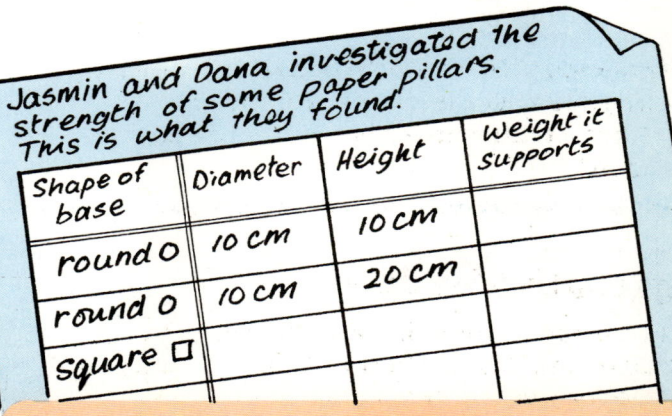

Jasmin and Dana investigated the strength of some paper pillars. This is what they found.

Shape of base	Diameter	Height	Weight it supports
round O	10 cm	10 cm	
round O	10 cm	20 cm	
square ▢			

Loads are sometimes supported on ARCHES

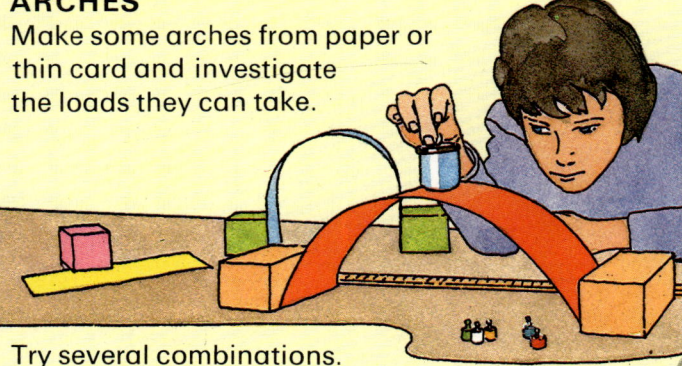

Make some arches from paper or thin card and investigate the loads they can take.

Try several combinations.

Same length of card; Vary distance of span.
Vary length of card; Keep same span.

Investigate different ways of supporting a road by an arch

- Can you think of different circumstances which would need the different bridge types?
- Does the material from which you make the bridge affect the decision?

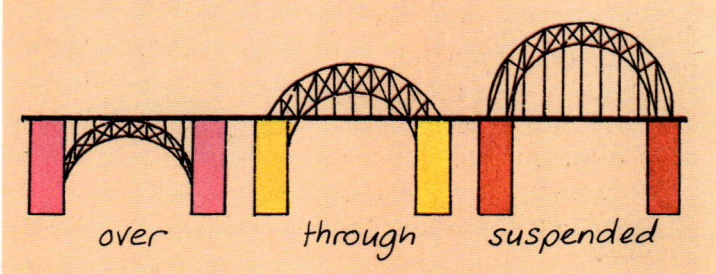

over   through   suspended

39

Bridges

Most of us live near a bridge. Think about the ones near you and ones that you can find out about from books or postcards.
- All bridges span a gap, but they have many different purposes. Consider what goes over and what goes under each bridge.
- Compare the bridges:
  – what materials they are made from;
  – the type of structure;
  – the position of each one.
- Is there any pattern in your findings? Does it relate to the bridge's age?

Exploration and investigation

Building a suspension bridge

To understand the forces involved, try this with a group of friends.

1. Lift a person or heavy load, like this. 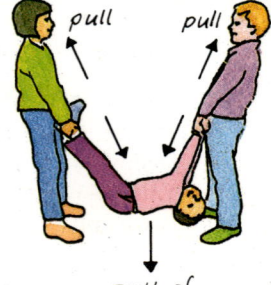
2. Now get two extra people to pull back on the 'support'. It is now easier to hold up the suspended weight.

Now you are ready to build your suspension bridge.

1. Build your towers and anchor them. 
2. Put in the suspension cable.
3. Suspend the road deck from the cable. The cable takes the weight of the road and the pillars support it.

Challenge 1

Can you build a structure which will hold a weight out as far from the main pillar as possible?

How important is the height of the tower?

How important is the support at the anchor point?

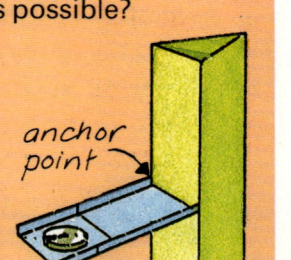

You could make this using straws, construction kits, or strips of card.

REMEMBER   Use your knowledge of forces and strong shapes to help you.

Challenge 2

Can you now use two structures from Challenge 1 and instead of a weight put a piece of card or wood to overlap and make a bridge?

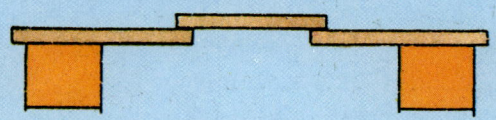

How wide a gap can you span with your bridge? When you reach the limit, see where the weak points are.

Can you modify it to take a heavier load?

This is a cantilever bridge.

Bridges and tunnels

Over & under

keystone

For a long time arch bridges have been used to span a river and carry a road on top.

The forces on an arch and the part played by the keystone are very important. If you can make your own arch bridge and test it, you will find you learn a lot.

Exploration and investigation

As a load moves across a bridge, the forces on any one point will vary. Bridge makers (designers and engineers) need to take this into consideration.

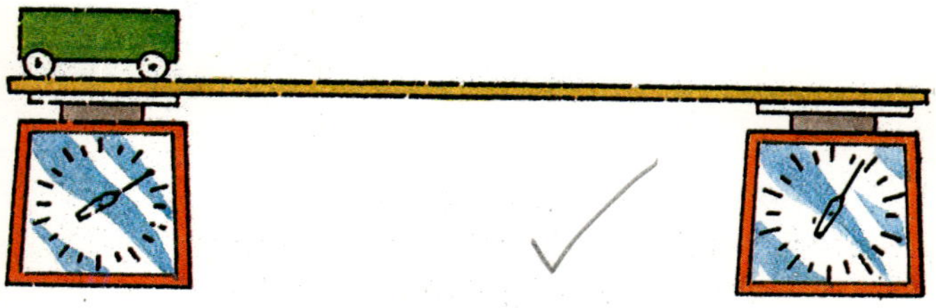

As you move the load across the 'bridge', watch how the force is shared between the two scales.

Investigate the forces involved in a load moving across a bridge

- Using the two scales as the pillars of the bridge, record the forces as the weight moves along the bridge:

Distance from end	Force on ① LHS balance	Force on ② RHS balance	① + ②
5 cm	4.4 Newtons	1 Newton	5.4 N
10 cm	4.2 N	1.2 N	5.4 N
15 cm			

- Make your road two lane and have two vehicles, one from either end, approaching each other and passing. How does this affect the forces on the bridge?

Tunnels are built under water, under land or through mountains, to take road and rail traffic. The water or earth above them exerts a very large force and this has to be considered by the designer/engineer.

Never dig holes or tunnels and crawl into them! They could collapse on you.

Investigating the best shape for a tunnel

Make some tubes about 20 cm long and about 10 cm diameter out of paper or thin card.
- Bury them in sand with one end open so that you can see what is happening.
- Pile up the sand on top.
- Which shaped tunnel could withstand most pressure from above? 
- Where would the engineer need to strengthen his or her design?

Houses

Houses

Building a house is a skilled operation. Creating a structure that provides a safe, weatherproof shelter that will last, is important.

Make a survey of all the different types of house in your area. [Your could enter this on a database.]

Look at and record:
- materials used
- type of house
- windows
- roof.

Notice the basic shapes of walls, roofs, windows, doors and chimneys.

Exploration and investigation

- Make some box-shape frames that are similar.

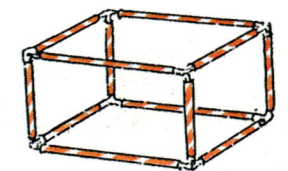

Try using straws, thin wood or card strips.
- How strong is your frame? Test one to find out. Will it support a mass?
- Where are the weakest points? Can you find ways to make them stronger?

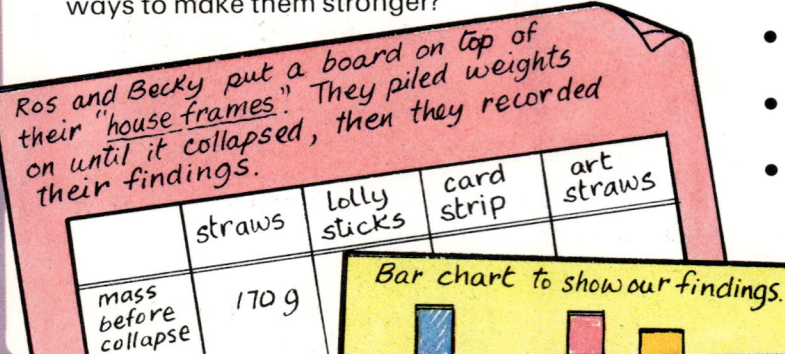

Ros and Becky put a board on top of their "house frames". They piled weights on until it collapsed, then they recorded their findings.

	straws	lolly sticks	card strip	art straws
mass before collapse	170 g			

Bar chart to show our findings.

Roofs

A roof is normally a triangular prism.

- Make some roof structures using the same materials as for the box-shape frames.
- What are the main forces which apply to a real roof?
- Devise a way to test your structure, bearing in mind the forces a roof is likely to experience.
- Try strengthening them with struts where there are weak points.
- Construction kits can be very useful for trying out ideas.

Windows and doorways are strengthened over the top with **lintels**. These are made of metal or reinforced concrete.
- Use model bricks or blocks to build a wall and take out some bricks to form a door hole.
- Compare the way the wall stands up to pressure with and without a lintel in position.

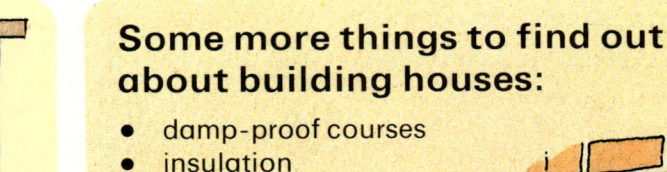

Some more things to find out about building houses:
- damp-proof courses
- insulation
- keeping things level as you build.

Towers

Towers

Look closely at the towers near you, including lampposts and electricity pylons.
- What materials are they made from?
- What are they used for?
- Why do they need to be tall?
- Towers need to stand up to wind and rain, or snow. Can you see signs showing how the designers have already thought of this?

Exploration and investigation

Build some towers

- Use:
  – straws and pins
  – straws and pipe cleaners
  – newspaper
  – strips of paper, card etc.
  – as many different types of construction kits as you can.
- Try to build the tallest tower you can. (It must be free standing.)
- You should be able to make it over 1 metre high. Then you should look for the weak points and try to strengthen them.

Investigating paper towers and how they stand up to winds

- Make a series of paper towers similar in size but with different cross-sections.
- Use a hair dryer to create a wind. Which ones topple first? 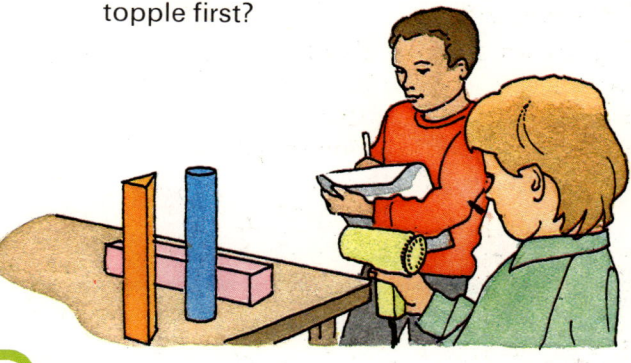
- Does it matter which side of the tower faces the fan? 
- Cut slots in your tower at 5-cm intervals.
- Push a piece of plasticine into each hole in turn and measure how close the fan can be before the tower topples over. 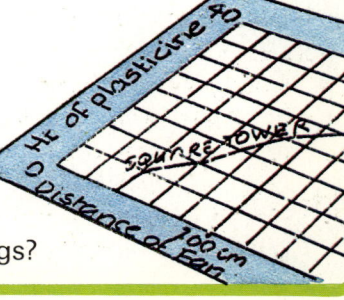
- Record your results.
- Repeat with other shaped towers.
- Is there a pattern in your findings?

Buttresses

- Look again at the examples of towers, lampposts and tree trunks.
- Can you see how the structures are stabilised?
- As you make your own towers, add supports or buttresses and see if it makes them more stable.

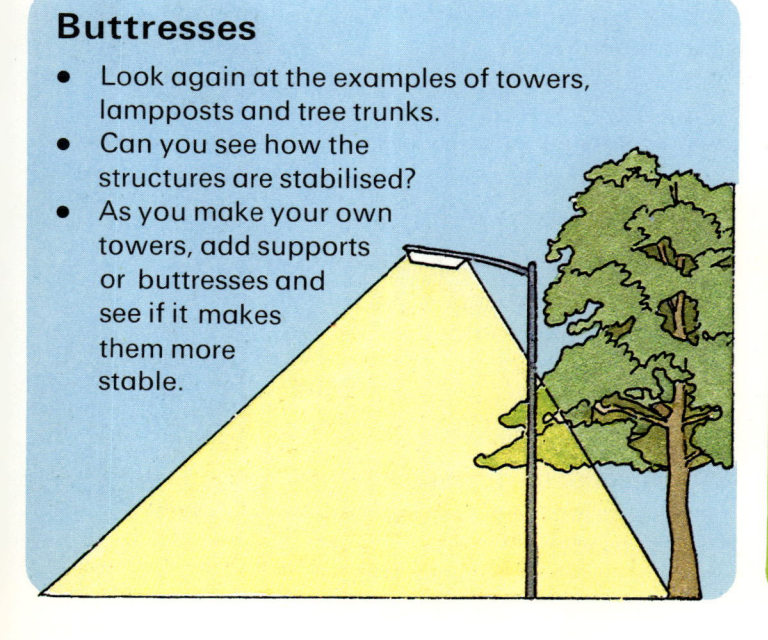

Cranes

Cranes

Finding out about cranes

A building site or docks are good places to observe cranes at work.
- How many different types can you find?
- Watch different movements.
- Look for strong shapes and list them.
- Notice which parts are solid and stable and which parts are more like a framework.
- Is there a reason for this?

> Building sites and docks can be dangerous places. Go with an adult.

Exploration and investigation

Moving loads
- Try lifting the load on its own.
- Now try with the plank over a chair.
- Feel the difference.

Investigating the most effective way of moving a heavy load
- Try moving the plank (pole) into different positions on the chair.
- Where is the best place to balance the plank?
- What happens to the balance if you change the position of the load?
- Which is easier for lifting – when you have a short end or a long end?

Did you know?

For thousands of years people have developed ways of lifting awkward or heavy loads. They have often used this 'lever' principle.

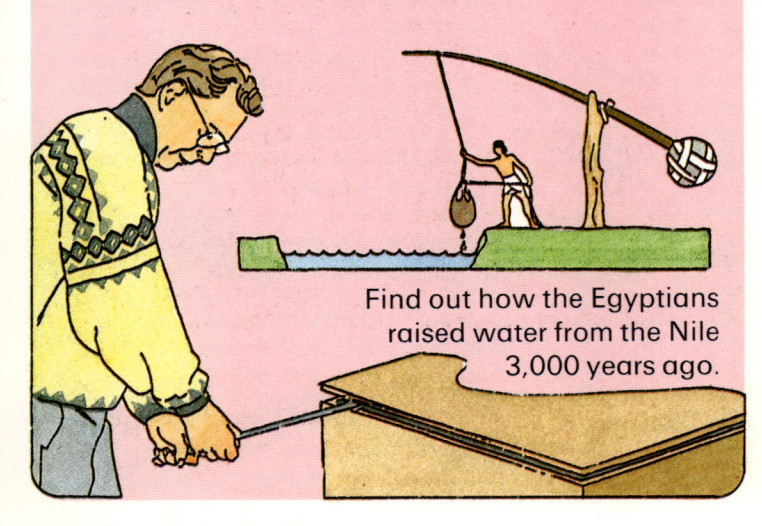

Find out how the Egyptians raised water from the Nile 3,000 years ago.

Challenge

Make a model which will lift a load up and down and will also swing it round. Use everyday materials or a construction kit.

- Is the model stable when it is loaded?
- Where is the balance point?
- Can you feel the force as you lift the load?

Think about the forces involved. It should help you balance them and build a stronger, more stable model.

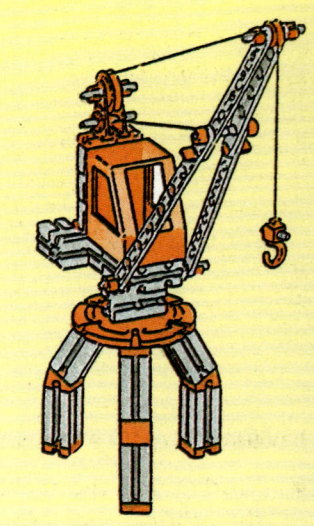

Cranes

Simple cranes

- Notice the mechanism at each end of the jib. Can you improve it?
- Choose the best thread.
- Can you see how the work is made easier?

These mechanisms could be useful.

Make a model of a simple jib crane that will allow you to raise and lower loads.

Exploration and investigation

Explore ways of raising and lowering loads safely

- Can you stop the load part way?
- Adapt your crane so that it will transfer the load to another place.
- What kind of movement is needed:
  - when the crane is fixed?
  - when the crane is mobile?

Can you make the angle of the jib variable?

Investigate the stability of the crane

- Devise ways of testing your crane under different conditions.
  - Try a range of load weights.
  - Vary the length of the jib.
  - Alter the angle of the jib.

Modify your crane as a result of your findings.

Record what you have done, your modifications and why you made them.

- Use meccano or a similar kit to make some working cranes.

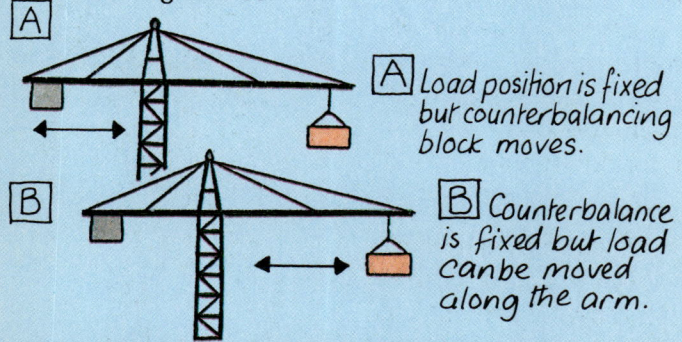

  A Load position is fixed but counterbalancing block moves.

  B Counterbalance is fixed but load can be moved along the arm.

Investigate the loads the cranes will carry

- Predict and then test the best position to get the balance right for a range of loads.
- Is there a pattern? If you double the load, do you halve the distance along the arm to balance it?

Alvin, Tim and Dinesh made a crane

Take care when testing with heavy loads. Work near the floor and protect toes!

Challenges

Here are some more challenges to set you thinking and doing. There is no right answer. For each one you will need some knowledge which you can combine with your own ideas. You will soon find yourself setting your own challenges. If your first ideas do not work, don't despair – talk to others, including your teacher. – Happy problem solving!

1. Design and make an end-pylon from straws which will withstand a pull of 1 Newton (100 g) from one direction only.

2. Design and make a model Big Wheel for a fairground. Remember your passengers must stay upright all the time!

3. Design and make a raft from newspaper which will hold 20 marbles. If you made it out of wood (lolly sticks) or straws, how would you change it?

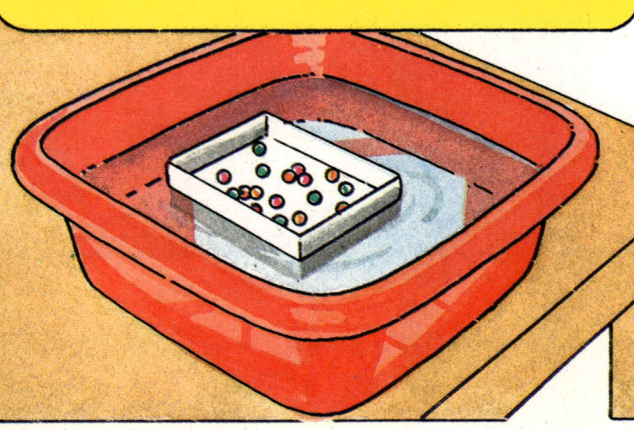

4. You need a crane which will pick things up from one end of the dock and put them down at the other end. Make a mobile crane to do this.

Challenges

5 Can you make a container which will hold and protect an egg even when it is dropped from 2 metres?

6 You need to cross a river which is not very wide but is deep enough for large ships to pass through. The surrounding land is flat. Can you design a suitable model bridge?

7 Design and make a model drawbridge for a castle using a mechanism which will allow it to be raised and lowered smoothly. Can you make it automatic?

8 Build a tower which has a platform on the top; so that a reporter can see and commentate on a race. (Remember to protect the reporter from bad weather. Have you thought how he/she will get up and down?)

Index

alarms 5, 23
arches 39
balance 28–9
bicycles 26
bonding materials 36
bridges 39–41, 47
buildings 40–2
buttresses 43
candles 2, 23
challenges 22–3, 46–7
colours 10, 12–13, 23
communication 2–23
containers 32, 47
cranes 44–6
ears 14
electricity 4–5
eyes 10
fastenings 34–5
fibres 37
forces 24–31, 41
friction 25, 30
gravity 25
hearing 14–17
houses 42
insulation 14, 33
lenses 11
levers 27, 44
light 2–13, 22
lintels 42
listening 14–21
magnifiers 11
materials 32–3
messages 4, 21, 23
metals 33
mirrors 8–9, 22
mortars 36

opaque 6
percussion instruments 17
pillars 39
pitch 17–18
plastics 33
reflections 8–9, 13
ropes 37
senses 2
shadows 7
shape 38
sight 10–11
signs 13
sounds 14–23
spectacles 11
stability 29, 33, 43–5
stethoscope 20
streamlining 31
strength 36–9
structures 24–47
sundials 8
torches 3, 6, 9, 12, 22
towers 40–3, 46–7
translucency 6–7
transparency 6, 33
tunnels 41
vibration 16, 18, 21
volume 17
water wheel 30
wind 31
wind instruments 19
wood 33, 35